Springer Tracts in Modern Physics
Volume 219

Springer Tracts in Modern Physics

Springer Tracts in Modern Physics provides comprehensive and critical reviews of topics of current interest in physics. The following fields are emphasized: elementary particle physics, solid-state physics, complex systems, and fundamental astrophysics.
Suitable reviews of other fields can also be accepted. The editors encourage prospective authors to correspond with them in advance of submitting an article. For reviews of topics belonging to the above mentioned fields, they should address the responsible editor, otherwise the managing editor.
See also springer.com

Christian Schüller

Inelastic Light Scattering of Semiconductor Nanostructures

Fundamentals and Recent Advances

With 105 Figures

Springer

Christian Schüller
Physik II
Institut für Experimentelle
und Angewandte Physik
Universität Regensburg
Universitätsstr. 31
93053 Regensburg
Germany
Email: christian.schueller@physik.uni-regensburg.de

Library of Congress Control Number: 2006930613

Physics and Astronomy Classification Scheme (PACS):
78.30.-j, 68.65.-k, 73.21-b, 78.67.-n, 81.07.-b

ISSN print edition: 0081-3869
ISSN electronic edition: 1615-0430
ISBN-10 3-540-36525-7 Springer Berlin Heidelberg New York
ISBN-13 978-3-540-36525-9 Springer Berlin Heidelberg New York

Typesetting: by the author using a Springer LaTeX macro package
Cover concept: eStudio Calamar Steinen
Cover production: *WMXDesign* GmbH, Heidelberg

Printed on acid-free paper SPIN: 10975671 56/techbooks 5 4 3 2 1 0

To Claudia, Sarah, Alina,

and my dear father, Hans Schüller,
who passed away during
preparation of this book.

Preface

Semiconductor nanostructures are currently one of the largest and most exciting areas in solid state physics. Low-dimensional electron systems (realized in semiconductor quantum structures) are particularly appealing because they allow one to study many-particle effects in reduced dimensions. Inelastic light scattering gives direct access to the elementary excitations of those systems. After an overview of the basic concepts and fabrication techniques for nanostructures on an introductory level, and an introduction into the method of inelastic light scattering, this monograph presents a collection of recent advances in the investigation of electronic elementary excitations in semiconductor nanostructures. Experiments on quantum wells, quantum wires, and quasiatomic structures, realized in quantum dots, are discussed. Theories are presented to explain the experimental results. Special chapters are also devoted to recent developments concerning tunneling – coupled systems and nanostructures embedded inside semiconductor microcavities. I have tried to make the chapters as self-containing as possible so that readers who are already familiar with the basics can directly read selected chapters.

With this book I have tried to fill the gap between research articles and contributed book chapters on special topics of the field on one hand, and more standard semiconductor textbooks (which cover a much broader range) on the other hand. The book should therefore be interesting for experimentalists, theorists, and research students working in the field of semiconductor nanostructures, as well as for graduate students with knowledge in solid state physics and quantum mechanics.

Most of the experimental and theoretical results presented in this book comprise a good part of the research that we have done at the Institute of Applied Physics and Microstructure Research Center of the University of Hamburg during the past decade. This work was only possible due to the collaboration with many excellent Diploma and Ph.D students. It is with pleasure that I thank Dr. Gernot Biese, Katharina Keller, Dr. Roman Krahne, Dr. Edzard Ulrichs, Dr. Lucia Rolf, Dr. Tobias Kipp, Dr. Maik-Thomas Bootsmann, Thomas Brocke, Gerwin Chilla, and Dr. Annelene Dethlefsen for an excellent and enjoyable collaboration in the Raman laboratory. Special thanks go to Professor Dirk Grundler and Professor Can-Ming Hu, my fellow postdocs in the Hamburg group, for many inspiring discussions. Very special thanks, however, go to Professor Detlef Heitmann, my mentor during my time in Hamburg. Among all my scientific teachers, he had by far the greatest impact on my scientific life and career. Our work immensely benefited from his enthusiasm and deep knowledge and I appreciate the many lively discussions which took place in a very friendly and convenient atmosphere.

Regensburg
May 2006

Christian Schüller

Contents

1 **Introduction** ... 1
References ... 4

Part I Basic Concepts

2 **Fundamentals of Semiconductors and Nanostructures** ... 9
2.1 III-V Semiconductors: Crystal and Band Structure ... 9
2.1.1 Phenomenology ... 9
2.1.2 k∗p Theory ... 13
2.2 Electrons in Three, Two, One, and Zero Dimensions ... 16
2.3 Layered Growth of Semiconductors: Vertical Nanostructures . 18
2.3.1 Molecular–Beam Epitaxy (MBE) ... 19
2.4 Electronic Ground State of Vertical Nanostructures ... 22
2.4.1 Envelope Function Approximation (EFA) ... 22
2.4.2 Self–Consistent Band Structure Calculation ... 25
2.5 Lateral Micro- and Nanostructures ... 30
2.5.1 General Remarks ... 30
2.5.2 Lithography and Etching ... 31
2.5.3 Self–Assembled Quantum Dots ... 35
2.6 Electronic Ground State of Lateral Nanostructures ... 37
References ... 38

3 **Electronic Elementary Excitations** ... 41
3.1 Single–Particle Continua ... 42
3.2 Electron–Density Waves: Phenomenology of Collective Charge– and Spin–Density Excitations ... 43
3.3 Collective Excitations: Theoretical Models ... 48
3.3.1 Basic Ideas of RPA and TDLDA ... 49
3.3.2 Application to Two–Subband System ... 50
3.3.3 Plasmon–LO Phonon Coupling ... 53
References ... 54

4 Basic Concepts of Inelastic Light Scattering, Experiments on Quantum Wells 57
4.1 Macroscopic Approach 57
4.1.1 General Remarks 57
4.1.2 Macroscopic Point of View 59
4.1.3 Dissipation–Fluctuation Analysis 61
4.2 Microscopic Approach, Polarization Selection Rules 62
4.2.1 Two- and Three-Step Scattering Processes 62
4.2.2 Scattering Cross Section: General Considerations 68
4.2.3 Scattering by Crystal Electrons: Polarization Selection Rules 71
4.2.4 Parity Selection Rules in Nanostructures 75
4.2.5 Intrasubband Excitations, Grating Coupler–Assisted Scattering 76
4.2.6 Multiple Cyclotron Resonance Excitations in Quantum Wells 79
References 83

Part II Recent Advances

5 Quantum Dots: Spectroscopy of Artificial Atoms 87
5.1 Introduction 87
5.2 Semiconductor Quantum Dots 90
5.2.1 Preparation of Quantum Dots 90
5.2.2 Electronic Ground State and Excitations 91
5.3 GaAs–AlGaAs Deep-Etched Quantum Dots 95
5.3.1 Parity Selection Rules in Quantum Dots 96
5.3.2 Fine Structure in Quantum Dots 98
5.3.3 The Important Role of Extreme Resonance 104
5.3.4 Calculations for Few-Electron Quantum Dots 109
5.4 InAs Self-Assembled Quantum Dots 112
5.4.1 Few–Electron Quantum–Dot Atoms 112
5.4.2 Electronic Excitations in InAs SAQD 113
5.4.3 Comparison with Exact Calculations 114
References 118

6 Quantum Wires: Interacting Quantum Liquids 121
6.1 Introduction 121
6.2 Electronic Elementary Excitations in Quantum Wires 122
6.2.1 Ground State and Excitations 122
6.2.2 Experimental Spectra and Wave–Vector Dependence 125
6.3 Confined and Propagating 1D Plasmons in a Magnetic Field 130
6.3.1 Microscopic Picture for Confined Plasmons 130
6.3.2 Coupling with Bernstein Modes 134

6.4 Towards the Tomonaga–Luttinger Liquid? 138
References 142

7 Tunneling–Coupled Systems 145
7.1 Introduction 145
7.2 Charge–Density Excitation Spectrum in Tunneling–Coupled Double Quantum Wells 146
7.3 Experiments on Tunable GaAs–AlGaAs Double Quantum Wells 150
7.4 Vertically–Coupled Quantum Wires 153
References 158

8 Inelastic Light Scattering in Microcavities 161
8.1 Introduction 161
8.2 2DES Inside a Semiconductor Microcavity 162
8.3 Optical Double–Resonance Experiments 163
References 168

Part III Appendix

Kronecker Products of Dipole Matrix Elements I 171

Kronecker Products of Dipole Matrix Elements II 173

Index 175

1 Introduction

Charge carriers in modulation–doped semiconductor quantum systems are a field of enormous and still growing research interest since they allow, in specially tailored systems, the investigation of fundamental properties, such as many–particle interactions, of electrons in reduced dimensions. Over the past decades, the experimental investigation of interacting electrons in low dimensions has led to many new and sometimes unexpected insights into many–particle physics in general. Famous examples are unique electronic transport properties as the integer and fractional quantum–Hall effects in quasi two–dimensional (Q2D) systems. Quasi one–dimensional (Q1D) electron systems, realized in semiconductor quantum wires, have been the subject of intense theoretical and experimental debates concerning the character – Fermi–liquid or Luttinger–liquid – of the interacting Q1D quantum liquid. During the past few years, tunneling–coupled electronic double–layer structures have been revisited as very interesting candidates for the realization of new quantum phases in an interacting many–particle system. A new quality came into the physics of semiconductor nanostructures by the development of quantum systems, embedded in microresonators, also called microcavities. This new inventions allowed one to investigate the light–matter interaction from an advanced point of view.

Optical spectroscopy techniques, like far–infrared (FIR) transmission [1–14] and resonant inelastic light scattering (or Raman) spectroscopy, are ideal tools to study the spectrum of elementary electronic excitations of those systems. Since the 1970's, inelastic light scattering has proven to be a very useful and powerful tool in the investigation of electrons or holes in semiconductors. Especially in the study of particle–particle interactions or coupling with other elementary excitations, inelastic light scattering experiments are extremely fruitful. In particular, also a finite quasimomentum q can be transferred to the excitations, which is in conventional backscattering geometry maximally that of the incoming laser light ($\approx 10^5$ cm^{-1}). The power of the method also results from the improvement of lasers and detectors in the visible and near–infrared spectral range where nowadays very powerful tunable lasers and detectors, such as charge–coupled–device cameras, are available. By the inelastic scattering of light, electronic elementary excitations with typical energies in the FIR spectral range can be observed in the visible range.

Christian Schüller: *Inelastic Light Scattering of Semiconductor Nanostructures*
STMP **219**, 1–5 (2006)
DOI 10.1007/3-540-36526-5_1

The first experiments of inelastic light scattering by free electrons were performed by Mooradian and Wright in 1968 [15], who studied collective plasma oscillations (plasmons), coupled to LO phonons, in n–type bulk GaAs. Later Mooradian also observed under resonant excitation, i.e., the laser frequency is in the vicinity of the optical $E_0 + \Delta$ energy gap of the semiconductor, excitations which – at that time – were interpreted as single–particle excitations [16]. According to the experimental findings, Hamilton and McWhorter deduced in their theoretical work that this single–particle scattering, which results from so called spin–density fluctuations, can be observed in Zincblende–type semiconductors in depolarized scattering geometry, i.e., the polarization directions of incoming and scattered light are perpendicular to each others. Scattering by plasmons due to charge–density fluctuations occurs in parallel polarization configuration (polarized geometry) [17].

In contrast to light scattering by optical phonons, the electronic Raman signals are strongly dependent on resonance enhancement effects at optical energy gaps. In many cases only these enhancement effects, which occur if the laser frequency is in the vicinity of such energy gaps, allow for the observation of electronic excitations. In 1978, E. Burstein proposed that, due to these resonance enhancements, light scattering should be sensitive enough to observe electronic excitations of Q2D electron gases with densities as low as 10^{11} cm^{-2} [18]. Such Q2D electron systems can be realized today in a nearly perfect way in modulation–doped GaAs–AlGaAs heterostructures or quantum wells, grown by molecular–beam epitaxy (MBE). Soon after this proposal, the first observations of Q2D intersubband excitations were reported by A. Pinczuk [19] and G. Abstreiter [20] in their pioneering works. In the following decade, a wealth of experiments on Q2D electron systems followed, which demonstrated the versatility of the resonant light scattering technique [21]. Through all the years it was commonly accepted that the electronic excitations, which can be observed by inelastic light scattering, fall into two main categories: Spin–density excitations (SDE) which were interpreted as single–particle excitations because exchange–correlation effects were assumed to be small (observed in depolarized geometry) and charge–density excitations (CDE, plasmons) which can be observed in polarized geometry (see, e.g., [21]). The latter are depolarization shifted with respect to the corresponding SDE due to direct Coulomb interaction [22]. Very surprisingly, in contradiction to this long lasting assumption, Pinczuk et al. demonstrated in another pioneering work in 1989 that in high–mobility quantum–well samples, *additionally* to the intersubband SDE and CDE, excitations can be observed with energies in between those of the SDE and CDE and which occur in *both* polarization configurations [23]. These excitations showed all features which one expects from pure single–particle excitations. This completely changed the point of view and from there on also SDE's were regarded as *collective* excitations of the electron gas, whereas the excitations which show no polarization selection rules were interpreted as *single–particle* excitations (SPE).

Theoretical considerations by state of the art calculations in the local–density [24, 25] or time–dependent Hartree Fock approximation [25] confirmed the experimentally observed *collective* shift of the SDE with respect to the SPE concerning the energetic positions. Nevertheless, there was so far no explanation for the existence itself of single–particle–like excitations in the Raman spectra. The calculations of Raman spectra, which for simplicity were made almost throughout for *nonresonant* conditions, exhibited no single–particle peak because these excitations are naturally screened by the interaction (see, e.g., [25]). In our work, we could show that under resonant scattering conditions, also from a theoretical point of view, excitations at single–particle level spacings can be expected. At the mean–field level, those excitations are single–particle excitations. Within an exact treatment, however, also the SPE are excitations of the interacting electron system, and hence subject to small but finite energy renormalization. Experimental and theoretical aspects will be discussed in Chap. 5.

The development of sophisticated structuring techniques in the 1990's allowed one to reduce the dimensionality further by the so called top–down approach and produce Q1D quantum wires and quasi–zero–dimensional (Q0D) quantum dots, starting from Q2D systems. The quantum dots can be regarded as some kind of artificial atoms [26]. In 1989 the first Raman experiments on electronic excitations in quantum wires were reported [27, 28]. Since then a number of papers appeared about, e.g., many–particle interactions and selection rules in those systems [29, 30, 31, 32, 33, 34] and investigations with applied external magnetic field [35, 36, 37]. In the past decade also first Raman experiments on Q0D quantum dots have been reported [29, 38, 39, 40, 41]. In particular the spectroscopy of self–assembled InGaAs quantum dots is very promising, since with these systems it is possible to study Q0D systems with only few electrons [41]. In tunneling–coupled systems, the interplay between Coulomb interaction and tunneling coupling can be investigated [42].

The book is divided into two main parts. In the first part, the basic concepts, which are necessary to follow the second part in detail, are presented and discussed. This comprises a brief introduction into the properties of semiconductors and their nanostructures (Chap. 2), the introduction into electronic elementary excitations (Chap. 3), and the principles of inelastic light scattering (Chap. 4). The second part of the book, where the recent advances in the field are summarized, consists of four chapters, devoted to the investigation of quantum dots (Chap. 5), quantum wires (Chap. 6), tunneling–coupled systems (Chap. 7), and to inelastic light scattering in microcavities (Chap. 8). Each chapter is written as self–containing as possible so that readers who are already familiar with the basics can directly read selected chapters. By doing so, it was not possible to completely avoid redundances but I tried to keep them as low as possible.

References

1. W. Hansen, M. Horst, J. P. Kotthaus, U. Merkt, Ch. Sikorski, and K. Ploog: Phys. Rev. Lett. **58**, 2586 (1987)
2. F. Brinkop, W. Hansen, J. P. Kotthaus, and K. Ploog: Phys. Rev. B **37**, 6547 (1988)
3. T. Demel, D. Heitmann, P. Grambow, and K. Ploog: Phys. Rev. B **38**, 12372 (1988)
4. T. Demel, D. Heitmann, P. Grambow, and K. Ploog: Phys. Rev. Lett. **66**, 2657 (1991)
5. H. Drexler, W. Hansen, J. P. Kotthaus, M. Holland, and S. P. Beaumont: Phys. Rev. B **46**, 12849 (1992)
6. M. A. Reed, J. N. Randall, R. J. Aggarwal, R. J. Matyi, T. M. Moore, and A. E. Wetsel: Phys. Rev. Lett. **60**, 535 (1988)
7. W. Hansen, T. P. Smith III, K. Y. Lee, J. A. Brum, C. M. Knoedler, J. M. Hong, and D. P. Kern: Phys. Rev. Lett. **62**, 2168 (1989)
8. C. Sikorski and U. Merkt: Phys. Rev. Lett. **62**, 2164 (1989)
9. T. Demel, D. Heitmann, P. Grambow, and K. Ploog: Phys. Rev. Lett. **64**, 788 (1990)
10. A. Lorke and J. P. Kotthaus: Phys. Rev. Lett. **64**, 2559 (1990)
11. B. Meurer, D. Heitmann, and K. Ploog: Phys. Rev. Lett. **68**, 1371 (1992)
12. K. Bollweg, T. Kurth, D. Heitmann, V. Gudmundsson, E. Vasiliadou, P. Grambow, and K. Eberl: Phys. Rev. Lett. **76**, 2774 (1996)
13. T. Darnhofer and U. Rössler: Phys. Rev. B **47**, 16020 (1993)
14. T. Darnhofer, M. Suhrke, and U. Rössler: Europhys. Lett. **35**, 591 (1996)
15. A. Mooradian and G. B. Wright: Phys. Rev. Lett. **16**, 999 (1966)
16. A. Mooradian: Phys. Rev. Lett. **20**, 1102 (1968)
17. D. Hamilton and A. L. McWhorter in: *Light Scattering Spectra of Solids*, ed. G. B. Wright (Springer, New York, 1969), p. 309
18. E. Burstein, A. Pinczuk, and S. Buchner in: *Physics of Semiconductors*, ed. B. L. H. Wilson (The Institute of Physics, London, 1979), p. 1231
19. A. Pinczuk, H. L. Störmer, R. Dingle, J. M. Worlock, W. Wiegmann, and A. C. Gossard: Solid State Commun. **32**, 1001 (1979)
20. G. Abstreiter and K. Ploog: Phys. Rev. Lett. **42**, 1308 (1979)
21. For an overview see: A. Pinczuk and G. Abstreiter in: *Light Scattering in Solids V*, Topics in Applied Physics Vol. 66, eds. M. Cardona and G. Güntherodt (Springer, Berlin, 1988) p. 153
22. T. Ando, A. B. Fowler, and F. Stern: Rev. Mod. Phys. **54**, 437 (1982)
23. A. Pinczuk, S. Schmitt–Rink, G. Danan, J. P. Valladares, L. N. Pfeiffer, and K. W. West: Phys. Rev. Lett. **63**, 1633 (1989)
24. T. Ando and S. Katayama: J. Phys. Soc. Jpn. **54**, 1615, (1985)
25. M. S.-C. Luo, S. L. Chuang, S. Schmitt–Rink, and A. Pinczuk: Phys. Rev. B **48**, 11086 (1993)
26. P. Maksym and T. Chakraborty: Phys. Rev. Lett. **65**, 108 (1990)
27. J. S. Weiner, G. Danan, A. Pinczuk, J. Valladares, L. N. Pfeiffer, and K. W. West: Phys. Rev. Lett. **63**, 1641 (1989)
28. T. Egeler, G. Abstreiter, G. Weimann, T. Demel, D. Heitmann, P. Grambow, and W. Schlapp: Phys. Rev. Lett. **65**, 1804 (1990)

29. C. Schüller, G. Biese, K. Keller, C. Steinebach, D. Heitmann, P. Grambow, and K. Eberl: Phys. Rev. B **54**, R17304 (1996)
30. G. Biese, C. Schüller, K. Keller, C. Steinebach, D. Heitmann, P. Grambow, and K. Eberl: Phys. Rev. B **53**, 9565 (1996)
31. A. R. Goñi, A. Pinczuk, J. S. Weiner, J. S. Calleja, B. S. Dennis, L. N. Pfeiffer, and K. W. West: Phys. Rev. Lett. **67**, 3298 (1991)
32. A. Schmeller, A. R. Goñi, A. Pinczuk, J. S. Weiner, J. S. Calleja, B. S. Dennis, L. N. Pfeiffer, and K. W. West Phys. Rev. B **49**, 14778 (1994)
33. C. Dahl, B. Jusserand, and B. Etienne: Phys. Rev. B **51**, 17211 (1995)
34. M. Sassetti and B. Kramer: Phys. Rev. Lett. **80**, 1485 (1998)
35. A. R. Goñi, A. Pinczuk, J. S. Weiner, B. S. Dennis, L. N. Pfeiffer, and K. W. West: Phys. Rev. Lett. **67**, 1151 (1993)
36. C. Steinebach, R. Krahne, G. Biese, C. Schüller, D. Heitmann, and K. Eberl: Phys. Rev. B **54**, R14281 (1996)
37. E. Ulrichs, G. Biese, C. Steinebach, C. Schüller, and D. Heitmann: Phys. Rev. B **56**, R12760 (1997)
38. R. Strenz, U. Bockelmann, F. Hirler, G. Abstreiter, G. Böhm, and G. Weimann: Phys. Rev. Lett. **73**, 3022 (1994)
39. D. J. Lockwood, P. Hawrylak, P. D. Wang, C. M. Sotomayor Torres, A. Pinczuk, and B. S. Dennis: Phys. Rev. Lett. **77**, 354 (1996)
40. C. Schüller, K. Keller, G. Biese, E. Ulrichs, L. Rolf, C. Steinebach, D. Heitmann, and K. Eberl: Phys. Rev. Lett. **80**, 2673 (1998)
41. T. Brocke, M.-T. Bootsmann, M. Tews, B. Wunsch, D. Pfannkuche, Ch. Heyn, W. Hansen, D. Heitmann, and C. Schüller: Phys. Rev. Lett. **91**, 257401 (2003)
42. M.-T. Bootsmann, C.-M. Hu, Ch. Heyn, D. Heitmann, and C. Schüller: Phys. Rev. B **67**, 121309(R) (2003)

Part I

Basic Concepts

2 Fundamentals of Semiconductors and Nanostructures

The majority of experiments of inelastic light scattering on semiconductor nanostructures has been performed on III–V semiconductors, like GaAs, as the most prominent example. In this chapter, an introduction into the basic properties of these materials is given. The first section gives a summary of the crystal and electronic band structure of the bulk material. After a short survey into the properties of electrons in different dimensions in the second section, growth methods for so called vertical nanostructures, i.e., layered heterostructures consisting of two different materials, are described in the third section. In these vertical nanostructures, quasi two–dimensional (Q2D) electron systems can be realized. This section is finalized by the description of commonly used concepts for theoretical calculations of the ground state of such systems. The second last section introduces the most important methods for the preparation of lateral micro and nanostructures. In those structures, the dimensionality of charge carriers or of quasi particles is reduced further by lithography and etching processes, or by self–organized growth methods, resulting in quasi one–dimensional (Q1D) or quasi zero–dimensional (Q0D) quantum structures. The section is finalized by an overview over methods for the calculation of the electronic ground state of lateral nanostructures. Readers who are already familiar with semiconductors and the fabrication and physics of nanostructures may skip this tutorial chapter and directly continue with Chap. 3.

2.1 III-V Semiconductors: Crystal and Band Structure

2.1.1 Phenomenology

Most III–V compound semiconductors, like GaAs, grow in Zincblende structure. The symmetry of this cubic lattice structure is described by the space group T_d^2. The corresponding point group, T_d, of the lattice sites is the symmetry group of the regular tetrahedron. It consists of 24 symmetry operations [1]. The Zincblende lattice is formed by two intersecting face–centered cubic (fcc) lattices, which are shifted by one quarter of the cubic space diagonal against each others. In Fig. 2.1, the spatial arrangement of Ga and As atoms in the Zincblende lattice is shown and compared to the diamond lattice (e.g.,

Christian Schüller: *Inelastic Light Scattering of Semiconductor Nanostructures*
STMP **219**, 9–39 (2006)
DOI 10.1007/3-540-36526-5_2

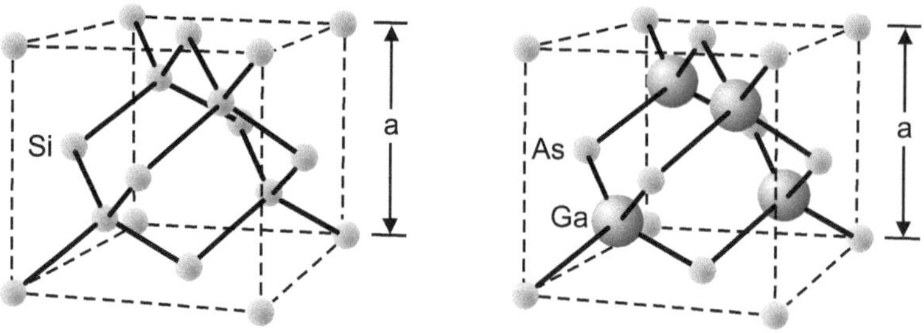

Fig. 2.1. Crystal structure of Silicon (*left*) and Galliumarsenide (*right*)

Si). The Zincblende lattice is no Bravais lattice, since its elementary cell contains two atoms, one at the origin and one at $(\frac{a}{4}, \frac{a}{4}, \frac{a}{4})$, where a is the lattice parameter. The reciprocal lattice of the fcc lattice, which is the underlaying lattice of the Zincblende structure, is a body–centered cubic (bcc) lattice. The Wigner–Seitz cell of the bcc lattice, which is the first Brillouin zone corresponding to the real space fcc lattice, is shown in Fig. 2.2. Some high–symmetry points, like the Γ– or the X–point, are indicated. Lattices of the point group T_d have no inversion symmetry, in contrast to semiconductors as, e.g., Si, which grow in the diamond structure (see Fig. 2.1).

The ternary alloy semiconductor $Al_xGa_{1-x}As$ is realized by replacing the fraction x of Ga atoms by Al atoms in the crystal lattice. Because of the statistical distribution of the atoms on the lattice sites of the Zincblende structure, the lattices of such ternary alloy semiconductors have no translational invariance. In principle, this has strong impact on the theoretical description of these structures, since electronic band structures, effective masses of electrons, etc., are no longer defined quantities. One usually circumvents these complications by introducing the so called virtual crystal approximation, which means that the real stochastic potential is replaced by an averaged potential which restores translational invariance. This guarantees that Bloch states, energy band gaps, and effective masses are defined. Usually with this

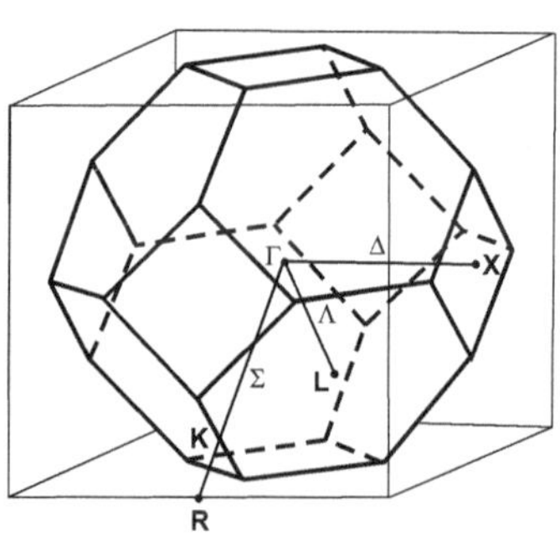

Fig. 2.2. First Brillouin zone of a face–centered cubic lattice

assumption, the empirical physical properties of the ternary alloys can be described quite well.

In the following we will introduce and discuss on an introductory level some common concepts of semiconductor physics, which are often used as the basis for the discussion of semiconductor nanostructures. We will start with an intuitive picture, which already gives us the most important features of the electronic band structure of III–V compound semiconductors. In those semiconductors, eight electrons per unit cell contribute to the chemical bonds between neighboring atoms. In a simplified picture, one can imagine that the s and p orbitals of neighboring atoms overlap and hybridize so that two new orbitals evolve: A bonding and an antibonding orbital. Since the crystal consists of a very large number of unit cells, the bonding and antibonding orbitals form bands. The bonding s orbitals have the lowest energies and are occupied with two electrons per unit cell. The remaining six electrons completely occupy the three bonding p orbitals. The bands which are formed by the antibonding orbitals are all unoccupied. The conduction band of the material is formed by the antibonding orbitals with lowest energy, the s band. Without spin–orbit coupling, the three p–like valence bands, which consist of the bonding p orbitals, are energetically degenerate at the Γ point. Figure 2.3 shows the bulk band structure of GaAs, calculated without spin–orbit interaction. At the Γ–point one can see the s–like conduction band, and the three–fold degenerate p–like valence band. As we will see in more detail below, the inclusion of spin–orbit coupling lifts the six–fold degeneracy of the valence band at the Γ-point: The p orbitals have an angular momentum quantum number of $L = 1$. If we add quantum mechanically the angular and spin quantum number to the total angular–momentum quantum number $\boldsymbol{J} = \boldsymbol{L}+\boldsymbol{S}$, we get the two possible values $J = |L+S| = 3/2$, and $J = |L-S| =$

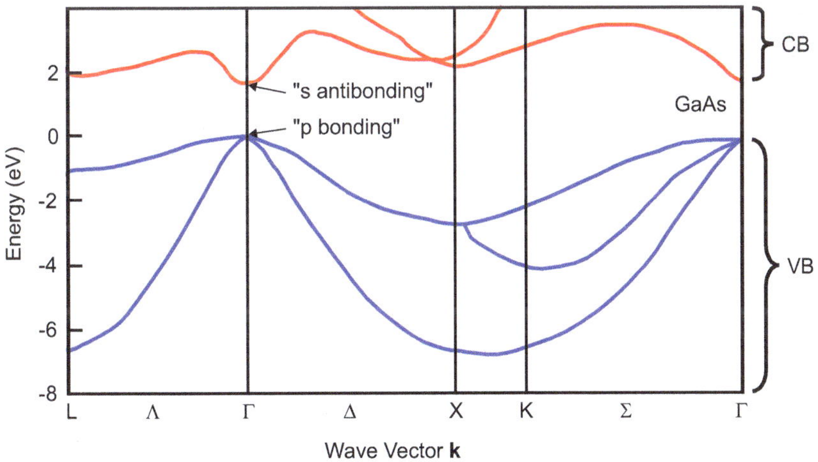

Fig. 2.3. Band structure of bulk GaAs, calculated without spin–orbit interaction

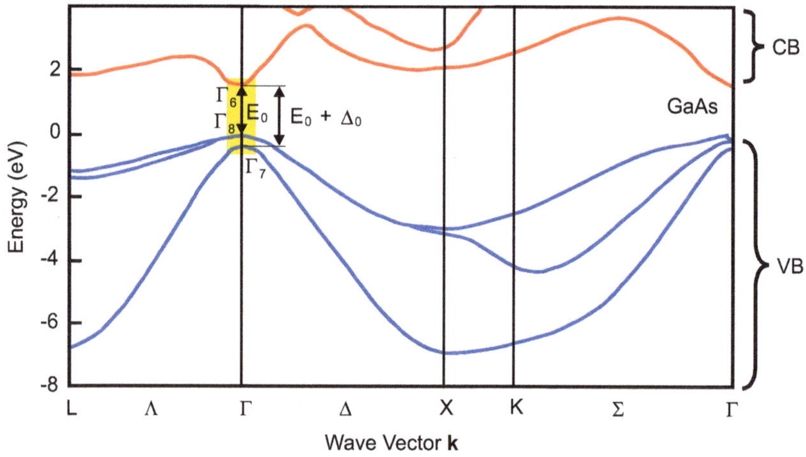

Fig. 2.4. Electronic bulk band structure of GaAs (data after [4])

1/2. The result is a quadruplet with Γ_8 symmetry ($J = 3/2$) and a doublet with Γ_7 symmetry ($J = 1/2$). This is displayed in Fig. 2.4, which shows the calculated band structure of GaAs, including spin–orbit interaction. The Γ_7 valence band is often called *split–off band.* The total angular momentum of the Γ_8 band is $J = 3/2$. Unoccupied states within this band with z–component of $J_z = \pm 3/2$ of the angular momentum are called heavy holes, and such with $J_z = \pm 1/2$ are called light holes. At finite wave vector $\boldsymbol{k}$ in the Brillouin zone, heavy and light holes split due to the reduced symmetry. However, for each of the bands – the heavy and the light holes – a two–fold spin degeneracy remains, if the lack of inversion symmetry of the lattice is neglected. The absence of this symmetry, e.g., in crystals of the point group T_d, or of lower symmetry, lead to a – mostly very small – lifting of the spin degeneracy, which is known as the *Dresselhaus effect* [2, 3]. For most of the inelastic light scattering experiments on free carriers in semiconductor nanostructures, and especially for all experiments which are considered in this book, only the Γ_6 conduction band and the Γ_8, and – in InGaAs material – the Γ_7 valence bands at the Γ–point are of relevance. This is so because to those points in $\boldsymbol{k}$ space, free electrons in the conduction band, or holes in the valence band, thermalize if they are injected either by doping or by photoexcitation. This area of further interest in the energy band diagram is highlighted in Fig. 2.4 by a vertical yellow bar.

GaAs is a semiconductor with a direct band gap, as can be seen from Figs. 2.3 and 2.4. $Al_x\ Ga_{1-x}As$ has a direct band gap for x values $x < 0.45$, only. For larger x, the band gap becomes indirect with a minimum at the X-point of the Brillouin zone.

We turn now to a more formal description of the electronic band structure. For calculation of the bulk band structure, one has to solve the one–electron

Schrödinger equation

$$\left[\frac{p^2}{2m_0} + V(\boldsymbol{r}) + \frac{\hbar}{4m_0^2c^2}(\boldsymbol{\sigma} \times \nabla V)\boldsymbol{p}\right]\psi(\boldsymbol{r}) = E\psi(\boldsymbol{r}) \,. \tag{2.1}$$

Here, m_0 is the free electron mass, $V(\boldsymbol{r})$ the periodic lattice potential, and $\boldsymbol{\sigma}$ the vector of the Pauli spin matrizes[1]. $V(\boldsymbol{r})$ contains an averaged electron–electron interaction and has the periodicity of the underlaying Bravais lattice. The third term on the left–hand side of (2.1) is a consequence of the spin–orbit interaction. The solutions of (2.1) are Bloch waves of the form

$$\psi_{n\boldsymbol{k}}(\boldsymbol{r}) = N u_{n\boldsymbol{k}}(\boldsymbol{r})e^{i\boldsymbol{k}\boldsymbol{r}} \,, \tag{2.2}$$

where N is a normalization factor and $u_{n\boldsymbol{k}}$ are lattice–periodic functions. A Bloch state $|n\boldsymbol{k}\rangle$ is characterized by a band index n and a crystal wave vector $\boldsymbol{k}$ of the first Brillouin zone of the reciprocal lattice. The explicit form of the Bloch states is in most cases not known. However, employing group theory, the behavior of the Bloch functions under the symmetry operations of the point group of the crystal lattice at high–symmetry points in the first Brillouin zone can be analyzed. In this way, one finds for the states, which correspond either to the conduction–band minimum or to the valence–band maximum at the Γ–point, the following four states: $|S\rangle$, $|X\rangle$, $|Y\rangle$, and $|Z\rangle$, whose wave functions transform like atomic s–, x–, y–, and z–functions. Including spin, we have the band–edge Bloch functions $|S\uparrow\rangle$, $|S\downarrow\rangle$, $|X\uparrow\rangle$, $|X\downarrow\rangle$, $|Y\uparrow\rangle$, $|Y\downarrow\rangle$, $|Z\uparrow\rangle$, and $|Z\downarrow\rangle$. Since in III–V semiconductors the spin–orbit coupling can not be neglected, it is advantageous not to use the above 8 band–edge Bloch functions as a basis but rather form linear combinations such that the spin–orbit interaction becomes diagonal. In this new basis, the total angular momentum, $\boldsymbol{J} = \boldsymbol{L} + \boldsymbol{S}$, as well as its projection along the z direction, J_z, are diagonal. For the s band edge, the addition of $L = 0$ and $S = 1/2$ results in $J = 1/2$, only (Γ_6 symmetry). For the p band edge, $L = 1$ and $S = 1/2$ gives either $J = 3/2$ or $J = 1/2$. In III–V compound semiconductors, the quadruplet $J = 3/2$ (Γ_8 symmetry) is always higher in energy than the doublet $J = 1/2$ (Γ_7 symmetry). The energetic difference between the Γ_7 and the Γ_8 band is the above introduced spin–orbit splitting Δ_0 (see also Fig. 2.4). The basis functions for which the spin–orbit interaction is diagonal are listed in Table 2.1 below.

2.1.2 k*p Theory

For the interpretation of experimental results, in many cases the band structure in a small range of wave vectors $\boldsymbol{k}$, around a high–symmetry point $\boldsymbol{k_0}$

[1] The components of $\boldsymbol{\sigma}$ are the 3 Pauli matrizes $\sigma_x = \begin{pmatrix} 0 & 1 \\ 1 & 0 \end{pmatrix}$, $\sigma_y = \begin{pmatrix} 0 & -i \\ i & 0 \end{pmatrix}$, and $\sigma_z = \begin{pmatrix} 1 & 0 \\ 0 & -1 \end{pmatrix}$.

Table 2.1. Periodic parts of the band–edge Bloch functions of a Zincblende lattice

symmetry	$\vert J, J_z\rangle$	ψ_{J,J_z}
Γ_6	$\vert\frac{1}{2}, \frac{1}{2}\rangle$	$S\uparrow$
Γ_6	$\vert\frac{1}{2}, -\frac{1}{2}\rangle$	$S\downarrow$
Γ_8	$\vert\frac{3}{2}, \frac{3}{2}\rangle$	$\frac{1}{\sqrt{2}}(X+iY)\uparrow$
Γ_8	$\vert\frac{3}{2}, \frac{1}{2}\rangle$	$\frac{i}{\sqrt{6}}[(X+iY)\downarrow -2Z\uparrow]$
Γ_8	$\vert\frac{3}{2}, -\frac{1}{2}\rangle$	$\frac{1}{\sqrt{6}}[(X-iY)\uparrow +2Z\downarrow]$
Γ_8	$\vert\frac{3}{2}, -\frac{3}{2}\rangle$	$\frac{i}{\sqrt{2}}(X-iY)\downarrow$
Γ_7	$\vert\frac{1}{2}, \frac{1}{2}\rangle$	$\frac{1}{\sqrt{3}}[(X+iY)\downarrow +Z\uparrow]$
Γ_7	$\vert\frac{1}{2}, -\frac{1}{2}\rangle$	$\frac{i}{\sqrt{3}}[-(X-iY)\uparrow +Z\downarrow]$

in the Brillouin zone, is of interest, only. The reason is that free carriers in semiconductors are often localized in the vicinity of such points of high symmetry. For such problems, the so called k*p method provides an adequate tool for the band structure calculation, since it enables a local description of the band structure with quite high accuracy on the millielectronvolt energy range. The k*p method was developed, e.g., by J. M. Luttinger and W. Kohn [5, 6] in order to generalize the effective mass approximation for the description of degenerate bands, or bands with band extrema which are not at the center of the Brillouin zone. This generalization was necessary in order to be able to calculate the band structures of, e.g., Si or Ge [5]. This theory can for instance be constructed on the basis of the above introduced band–edge Bloch functions, which is known as the so called Kane model [7, 8, 9, 10]. For global calculations of the band structure, over the whole range of the Brillouin zone, mostly pseudopotential or tight–binding methods are used (e.g., [4]). The main ideas of the k*p theory are the following:

Inserting the Bloch ansatz (2.2) into (2.1), yields

$$\left\{\frac{p^2}{2m_0} + V(\boldsymbol{r}) + \frac{\hbar}{4m_0^2c^2}(\boldsymbol{\sigma}\times\nabla V)\boldsymbol{p} + \frac{\hbar^2k^2}{2m_0} + \right.$$
$$\left. + \frac{\hbar\boldsymbol{k}}{m_0}\left(\boldsymbol{p} + \frac{\hbar}{4m_0c^2}\boldsymbol{\sigma}\times\nabla V\right)\right\} u_{n\boldsymbol{k}} = E_{n\boldsymbol{k}}u_{n\boldsymbol{k}}, \quad (2.3)$$

which is an equation for the $u_{n\boldsymbol{k}}$. Equation (2.3) can be written as

$$[H(\boldsymbol{k}=0) + W(\boldsymbol{k})]\, u_{n\boldsymbol{k}} = E_{n\boldsymbol{k}}u_{n\boldsymbol{k}}\ , \quad (2.4)$$

where $H(\boldsymbol{k}=0)$ is nothing but the Hamiltonian for $\boldsymbol{k}=0$ with eigenfunctions u_{n0}

$$H(\boldsymbol{k}=0)u_{n0} = E_{n0}u_{n0} \; . \tag{2.5}$$

Since the u_{n0} represent a complete set, one can write

$$u_{n\boldsymbol{k}} = \sum_m c_m(\boldsymbol{k})u_{m0} \; . \tag{2.6}$$

Inserting (2.6) into (2.4), multiplying from left with u^*_{m0}, and integrating over a unit cell delivers

$$\begin{aligned} \textstyle\sum_m & \left\{ \left(E_{n0} - E_{n\boldsymbol{k}} + \frac{\hbar^2 k^2}{2m_0} \right) \delta_{nm} + \right. \\ + & \left. \frac{\hbar \boldsymbol{k}}{m_0} \langle n0 | \boldsymbol{p} + \frac{\hbar}{4m_0c^2} (\boldsymbol{\sigma} \times \nabla V) | m0 \rangle \right\} c_m(\boldsymbol{k}) = 0 \; , \end{aligned} \tag{2.7}$$

where,

$$\langle n0|A|m0\rangle = A_{nm} = \int_{UC} u^*_{n0} \, A \, u_{m0} \, d^3r \; . \tag{2.8}$$

Equation (2.7) is equivalent to (2.4) and well suited for a perturbation approach in $\boldsymbol{k}$. For small $\boldsymbol{k}$ values, in the vicinity of the Γ–point, (2.7) yields a parabolic dispersion relation for nondegenerate bands (except spin degeneracy) [10]:

$$E_{n\boldsymbol{k}} = E_{n0} + \frac{\hbar^2}{2} \sum_{\alpha,\beta} k_\alpha \frac{1}{\mu_n^{\alpha\beta}} k_\beta \; ; \quad \alpha, \beta = x, y, z \; , \tag{2.9}$$

where

$$\frac{1}{\mu_n^{\alpha\beta}} = \frac{1}{m_0}\delta_{\alpha\beta} + \frac{2}{m_0^2} \sum_{m \neq n} \frac{\Pi^\alpha_{mn} \Pi^\beta_{nm}}{E_{n0} - E_{m0}} \; , \tag{2.10}$$

$$\boldsymbol{\Pi} = \boldsymbol{p} + \frac{\hbar}{4m_0c^2} (\boldsymbol{\sigma} \times \nabla V) \; . \tag{2.11}$$

For the description of the conduction band in the analysis of experiments on GaAs structures, this parabolic $E(\boldsymbol{k})$ relation with an isotropic effective mass $\mu_n^{\alpha\beta} \equiv m^* = 0.068 \; m_0$ [11] is used very often.

Calculations of the energetic dispersion of the valence bands is more involved, since here, e.g., the Γ_8 band is degenerate at the Γ–point. The simplest approach to calculate the dispersion of the heavy and light holes is to employ the above defined band–edge Bloch functions and neglect coupling to the conduction band and split–off valence band. Doing this within the k*p framework, one gets a 4×4 matrix for the Hamiltonian, which describes the kinetic energy of the heavy and light holes, in the vicinity of the center of the Brillouin zone [5]:

$$H_{\Gamma_8}(\boldsymbol{k}) = -\frac{\hbar^2}{2m_0}\begin{pmatrix} A_{3/2} & B & C & 0 \\ B^* & A_{1/2} & 0 & C \\ C^* & 0 & A_{-1/2} & -B \\ 0 & C^* & -B^* & A_{-3/2} \end{pmatrix} , \tag{2.12}$$

where

$$A_{\pm 3/2} = (\gamma_1 + \gamma_2)(k_x^2 + k_y^2) + (\gamma_1 - 2\gamma_2)k_z^2 , \tag{2.13}$$

$$A_{\pm 1/2} = (\gamma_1 - \gamma_2)(k_x^2 + k_y^2) + (\gamma_1 + 2\gamma_2)k_z^2 , \tag{2.14}$$

$$B = \sqrt{3}\gamma_3 k_z(k_x - ik_y) , \tag{2.15}$$

$$C = \frac{\sqrt{3}}{2}\left[(\gamma_2 - \gamma_3)(k_x + ik_y)^2 + (\gamma_2 + \gamma_3)(k_x - ik_y)^2\right] . \tag{2.16}$$

γ_1, γ_2, and γ_3 are so called Luttinger parameters[2]. With the Schrödinger equation

$$H_{\Gamma_8}(\boldsymbol{k})\psi_{\Gamma_8}(\boldsymbol{k}) = E(\boldsymbol{k})\psi_{\Gamma_8}(\boldsymbol{k}) \tag{2.17}$$

one needs to solve a system of 4 coupled differential equations to determine the electronic bulk band structure of the Γ_8 valence band in the vicinity of the center of the Brillouin zone. The wave functions $\psi_{\Gamma_8}(\boldsymbol{k})$ are four–component spinors in the basis of the 4 Γ_8 band–edge Bloch functions (see Table 2.1). In a more rigorous treatment, which shall not be described here, one can include also coupling to the conduction band (Γ_6 band) and to the spin–orbit split–off valence band (Γ_7 band). This results in an 8×8 Matrix for the Hamiltonian of the system in the vicinity of the Γ–point (see, e.g., the monograph [3]).

2.2 Electrons in Three, Two, One, and Zero Dimensions

An important feature of a quantum mechanical object – like the electron – is its density of allowed states. This density of states is strikingly different for electrons in three, two, one, or zero dimensions, as we will see below. Before we dive into the physics and technology of semiconductor nanostructures, we consider here the textbook example of a single electron in a box–like potential for introducing low–dimensional electron systems.

The well–known Schrödinger equation of an electron, moving in a potential $V(\boldsymbol{r})$, is

$$\left\{-\frac{\hbar^2}{2m_0}\nabla^2 + V(\boldsymbol{r})\right\}\psi_{lmn}(\boldsymbol{r}) = E_{lmn}\,\psi_{lmn}(\boldsymbol{r}) . \tag{2.18}$$

If we consider for $V(\boldsymbol{r})$ a box with sides L_1, L_2, and L_3, where the potential is zero inside, and infinitely high outside, $\psi_{lmn}(\boldsymbol{r})$ has to be zero at the boundaries. The analytic solutions of (2.18) are

[2] There is a variety of different sets for these parameters in literature. A commonly used set is, e.g. [12]: $\gamma_1 = 6.85$, $\gamma_2 = 2.1$, and $\gamma_3 = 2.9$.

$$\psi_{lmn}(\boldsymbol{r}) = \sqrt{\frac{8}{L_1 L_2 L_3}} \sin\left(\frac{l\pi x}{L_1}\right) \sin\left(\frac{m\pi y}{L_2}\right) \sin\left(\frac{n\pi z}{L_3}\right) , \tag{2.19}$$

with energy eigenvalues

$$E_{lmn} = \frac{\hbar^2\pi^2}{2m_0}\left[\left(\frac{l}{L_1}\right)^2 + \left(\frac{m}{L_2}\right)^2 + \left(\frac{n}{L_3}\right)^2\right] . \tag{2.20}$$

For a macroscopic solid, the lengths L_1, L_2, and L_3 are very large – in the range of millimeters – and the quasi momentum, $\boldsymbol{k} = (l\pi/L_1, m\pi/L_2, n\pi/L_3)$, is quasi continuous, leading to eigenenergies

$$E_{lmn} = E(\boldsymbol{k}) = \frac{\hbar^2 \boldsymbol{k}^2}{2m_0} . \tag{2.21}$$

In semiconductor nanostructures, some – or even all – of the L_i can be very small, in the range of nanometers. This leads to the situation that for small numbers of (l, m, n), the eigenenergies E_{lmn} can already be in the range of millielectronvolts, i.e., the far–infrared spectral range. The strict definition for a two–, one–, or zero–dimensional system is that one, two, or three of the L_i are exactly zero, respectively. However, in a real system, the L_i can not be exactly zero, but small. Therefore, the low–dimensional structures are called *quasi*–two, *quasi*–one, or *quasi*–zero dimensional, expressing their small but finite extension in certain spatial directions. With this, we can define for our simple electron–in–the–box system:

- Quasi zero–dimensional (Q0D) system: L_1, L_2, L_3 *small* $\longrightarrow$

$$E_{lmn} = \frac{\hbar^2\pi^2}{2m_0}\left[\left(\frac{l}{L_1}\right)^2 + \left(\frac{m}{L_2}\right)^2 + \left(\frac{n}{L_3}\right)^2\right] \tag{2.22}$$

  A Q0D system has a completely discrete energy spectrum.
- Quasi one–dimensional (Q1D) system: $L_1 \gg L_2$, L_3 $\longrightarrow$

$$E_{mn}(k_x) = \frac{\hbar^2 k_x^2}{2m_0} + \frac{\hbar^2\pi^2}{2m_0}\left[\left(\frac{m}{L_2}\right)^2 + \left(\frac{n}{L_3}\right)^2\right] \tag{2.23}$$

  In a Q1D system, the electron can move freely in one direction, only (x direction).
- Quasi two–dimensional (Q2D) system: L_1, $L_2 \gg L_3$ $\longrightarrow$

$$E_n(k_x, k_y) = \frac{\hbar^2 k_x^2}{2m_0} + \frac{\hbar^2 k_y^2}{2m_0} + \frac{\hbar^2\pi^2}{2m_0}\left(\frac{n}{L_3}\right)^2 \tag{2.24}$$

  In a Q2D system, the electron can move freely in two directions. Here, n is the index of the Q2D subbands.

Later in this book we will find that, when realized in semiconductor nanostructures, Q2D, Q1D, and Q0D systems are called *quantum wells*, *quantum wires*, and *quantum dots*, respectively. It shall be noted here that, in most of the experimentally realized structures, the lateral confining potentials are approximately parabolic, rather than box–like.

The density of states (DOS), $N(E)$, of a system is defined such that the quantity $N(E)\delta E$ is the number of solutions of the Schrödinger equation in the energy interval between E and $E+\delta E$. For three–dimensional, or strictly two–, or one–dimensional systems, $N(E)$ can be calculated by the expression

$$N(E) = \frac{dn(\boldsymbol{k})}{d\boldsymbol{k}} \frac{d\boldsymbol{k}}{dE} , \tag{2.25}$$

where $dn(\boldsymbol{k})/d\boldsymbol{k}$ is $(L_1)/(2\pi)$, $(L_1L_2)/(2\pi)^2$, or $(L_1L_2L_3)/(2\pi)^3$ for a one–, two–, or three–dimensional system, respectively. With this, one can find for the DOS, $n_{\mathrm{iD}}(E)$, of an i–dimensional system, normalized per unit volume, area, or length:

$$n_{\mathrm{3D}}(E) = \frac{1}{(2\pi)^3}\left(\frac{2\pi m_0}{\hbar^2}\right)^{3/2}\sqrt{E} , \tag{2.26}$$

$$n_{\mathrm{2D}}(E) = \frac{m_0}{2\pi\hbar^2} = const , \tag{2.27}$$

$$n_{\mathrm{1D}}(E) = \frac{1}{2\pi}\sqrt{\frac{m_0}{2\hbar^2}}\frac{1}{\sqrt{E}} . \tag{2.28}$$

The DOS of a zero–dimensional system is a delta function. Each solution of the Schrödinger equation can accommodate two electrons, one for each spin. This leads to an additional factor of 2 for the above densities of states. For a Q2D, Q1D, or Q0D system, we have of course separate branches of these densities of states for each *lateral* discrete energy. This means, in a Q2D or Q1D system, each 2D or 1D subband has its own 2D or 1D DOS, and for a Q0D system, each discrete energy eigenvalue has its own delta–like DOS. A schematic comparison of the various scenarios is collected in Fig. 2.5, which displays the momentum dispersions [Fig. 2.5(a)] and the DOS [Figs. 2.5(b)–(c)] of a 3D, as well as of strict and quasi 2D, 1D, and 0D systems. The particular shape of the DOS can have a profound influence on the transport and optical properties of low–dimensional electron systems. In particular, the spectrum of electronic elementary excitations, which can be probed in inelastic light scattering experiments, depends characteristically on the dimensionality, as we will see later.

2.3 Layered Growth of Semiconductors: Vertical Nanostructures

With sophisticated growth techniques, like molecular–beam epitaxy (MBE) or metal–organic chemical–vapor deposition (MOCVD), it is nowadays possi-

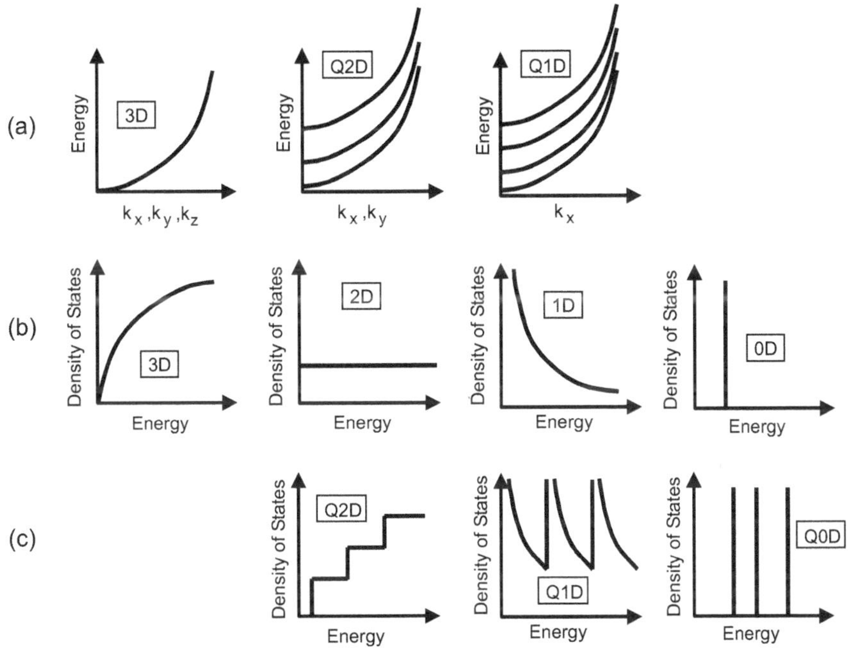

Fig. 2.5. Schematic pictures of (**a**) the energy dispersions of a 3D, Q2D, and Q1D electron, (**b**) the densities of states of a 3D, and strict 2D, 1D, and 0D electron, and, (**c**) the densities of states of a Q2D, Q1D, and Q0D electron

ble to prepare semiconductor multilayers, one atomic layer at a time. Thereby, one has independent control over the doping and composition in each layer. Such layered heterostructure samples are also called *vertical* nanostructures. We will here briefly describe the MBE technique[3], since with this technique high–quality semiconductor heterostructures, as used for inelastic light scattering experiments, are produced.

2.3.1 Molecular–Beam Epitaxy (MBE)

Figure 2.6 shows a schematic picture of an MBE machine. An MBE machine consists of an ultrahigh–vacuum chamber (10^{-14} mbar) with a diameter of approximately one meter. On one side of the chamber, a number of effusion cells are bolted onto the chamber. In these Knudsen cells, a refractory material boat contains a charge of one of the elemental species (e.g., Ga, Al, or As) for growth of the semiconductor, Si (for n–type doping), and Be or C (for p–type doping). Each boat is heated so that a vapor is obtained which leaves the cell for the growth chamber through a small opening. The vapor forms a beam that crosses the vacuum chamber to impinge on a substrate.

[3] For more details about the MBE and MOCVD techniques, see, e.g., [13, 14].

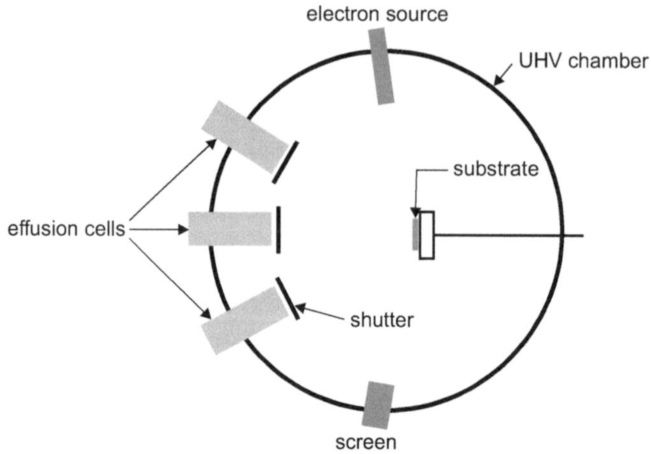

Fig. 2.6. Schematic picture of an MBE machine

For the growth of GaAs–AlGaAs structures, this substrate is usually GaAs. The flux rate is controlled by the temperature of the Knudsen cell. Shutters in front of the cells can be opened and closed within a time of about 0.1 s. This method enables one to grow a crystal, layer by layer, with a growth rate of about one monolayer per second.

Figure 2.7 displays the band gaps and lattice parameters of various III–V semiconductors. From this figure one can see, why the GaAs–AlGaAs system is almost perfectly suited for a heterostructure growth: Both materials have approximately the same lattice constant so that a perfect growth of one material on the other is guaranteed. The band gap energy, E_{Gap}, of $\mathrm{Al_xGa_{1-x}As}$ is larger than the band gap of GaAs. It depends linearly on the fraction x of Al atoms per unit cell [15]:

$$E_{Gap}(\mathrm{x}) = 1.5177 + 1.30\mathrm{x} \quad (\mathrm{in\,eV}) \,. \tag{2.29}$$

When grown epitaxially on each other, the band offsets between the two materials in the conduction and valence band are related like 70:30, respectively[4] [15, 17]. Thus, by the sequential growth of GaAs and $\mathrm{Al_xGa_{1-x}As}$ layers, heterostructures, like quantum wells or superlattices, can be realized with controlled doping (n– or p–type) in selected layers, and with monolayer precision. The most commonly used structures for state–of–the–art inelastic light scattering experiments are one–sided modulation–doped single quantum wells. Such a quantum well can be realized by growing a thin (typically 20 nm – 30 nm thick) GaAs layer in between $\mathrm{Al_xGa_{1-x}As}$ barriers (typical x values are around x= 0.33). The conduction– and valence–band edges in real space

[4] The commonly accepted relations, which can be found in literature, vary between about 70:30 and 60:40.

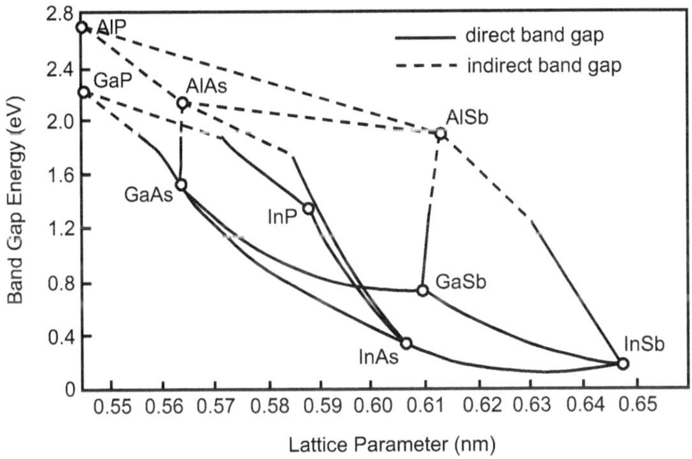

Fig. 2.7. Energy gaps and lattice parameters of III-V semiconductors (after [16])

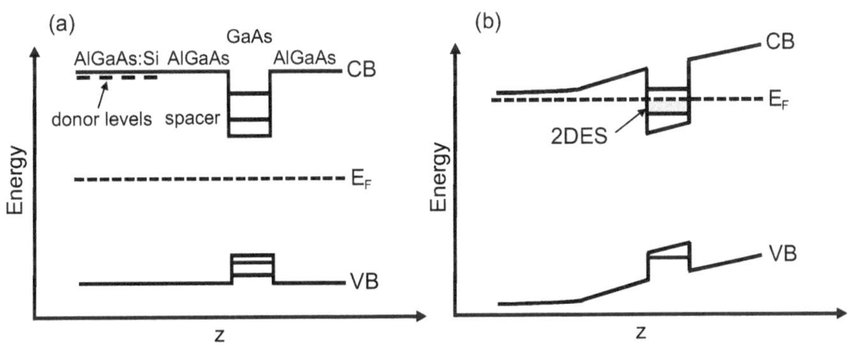

Fig. 2.8. Sketch of the conduction– and valence–band edges in real space of (**a**) an intrinsic GaAs–AlGaAs quantum well, and, (**b**) a one–sided modulation–doped structure

of an idealized, infinitely–large, intrinsic[5] single quantum–well structure are shown schematically in Fig. 2.8(a). Levels of neutral Si donors in the left AlGaAs barrier are indicated. This technique, of doping the barrier layers of a heterostructure, is called modulation doping [18]. Due to the potential discontinuities at the interfaces between the two materials, electrons are transferred from the donors to the GaAs quantum well and form a two–dimensional electron system (2DES) there[6]. The positive space charges of the ionized donors and the negatively charged electrons in the quantum well lead to a bending of

[5] GaAs bulk material has a p–type background doping due to residual carbon impurities.

[6] Because of the finite width of the quantum well, this is, of course, a Q2D electron system.

the potentials within the doped region and the well. This situation is shown in Fig. 2.8(b). In order to calculate the band structure and energy levels of such a heterostructure, one has to solve self–consistently the Schrödinger and Poisson equations of the structure (see, e.g., [19] and Sect. 2.4). The introduction of an undoped AlGaAs barrier layer (spacer) between the doped region and the quantum well leads to a larger separation of the ionized donors, which can act as scatterers, and the 2DES. This results in high electron mobilities in the range of 10^7 cm^2/Vs. In addition, also the one–sided doping, as displayed in Fig. 2.8, leads to higher electron mobilities than a symmetric two–sided doping. Due to the asymmetric potential shape, the electrons are confined more strongly to the left side of the well (see Fig. 2.8). Hence, the scattering due to interface imperfections is reduced as compared to symmetric wells, where the electrons feel two interfaces. GaAs–AlGaAs single quantum–well structures, as displayed in Fig. 2.8, are widely–used standard samples for inelastic light scattering experiments. They are also used as starting materials for the preparation of lower–dimensional structures, like quantum wires or quantum dots, by lithography and etching techniques (see Sect. 2.5.2).

2.4 Electronic Ground State of Vertical Nanostructures

In this section, the standard methods for describing the electronic ground state of vertical nanostructures – in particular of 2DES's – embedded in semiconductor heterostructures will be introduced. For such calculations, it is convenient to use the so called envelope–function approximation (EFA), which was originally developed by G. Bastard [10, 20]. Based on the EFA, the ground state of a doped heterostructure can be calculated by a self–consistent solution of the Schrödinger and Poisson equations of the semiconductor heterostructure.

2.4.1 Envelope Function Approximation (EFA)

The bulk wave function of an electron, as defined in the Bloch ansatz [see (2.2)], consists of two parts: A fast oscillating part, which has the periodicity of the lattice – the Bloch function – and a slowly varying part (as compared to the scale of the lattice parameter) – the plane wave. The plane–wave part is what is called the envelope function. In the EFA for semiconductor heterostructures, the envelope functions of different materials have to satisfy boundary conditions at the interfaces. The EFA is based on the 8 band–edge Bloch functions, which we introduced in Sect. 2.1. From this it follows that the method is only applicable to calculate the band structure in the vicinity of high–symmetry points in the Brillouin zones of the starting materials. The method works best for heterostructures, whose band edges are built up by the same band–edge Bloch functions in each material.

The main assumptions of the EFA are the following [10]:

1. In each layer, the wave function is expanded in a series of the periodic parts of the band–edge Bloch functions of the respective material.

$$\psi(\boldsymbol{r}) = \sum_l f_l^{(A)}(\boldsymbol{r})\, u_{l,0}^{(A)}(\boldsymbol{r}) \tag{2.30}$$

if $\boldsymbol{r}$ is in layer A , and

$$\psi(\boldsymbol{r}) = \sum_l f_l^{(B)}(\boldsymbol{r})\, u_{l,0}^{(B)}(\boldsymbol{r}) \tag{2.31}$$

if $\boldsymbol{r}$ is in layer B. The summation over l runs over all band edges, considered in the calculation (e.g., Γ_6, Γ_7, and Γ_8 in the case of an 8–band theory).

2. It is assumed that the periodic parts of the Bloch functions are the same in each layer:

$$u_{l,0}^{(A)}(\boldsymbol{r}) \equiv u_{l,0}^{(B)}(\boldsymbol{r}) \; . \tag{2.32}$$

With this assumptions, the heterostructure wave function can be written as

$$\psi(\boldsymbol{r}) = \sum_l f_l^{(A,B)}(\boldsymbol{r})\, u_{l,0}(\boldsymbol{r}) \; . \tag{2.33}$$

It follows that the heterostructure wave function is a product of fast oscillating functions – the $u_{l,0}$ –, which have the same periodicity as the lattices of the bulk crystals, and of slowly oscillating envelope functions f_l. The task is now to determine the functions $f_l^{(A,B)}(\boldsymbol{r})$. Because of the translational invariance of the point lattices in x–y–directions, the $f_l^{(A,B)}(\boldsymbol{r})$ can be factorized:

$$f_l^{(A,B)}(\boldsymbol{r}_{||}, z) = \frac{1}{\sqrt{S}}\, e^{i\boldsymbol{k}_{||}\boldsymbol{r}_{||}}\, \chi_l^{(A,B)}(z) \; , \tag{2.34}$$

where S is the sample area, and $\boldsymbol{k}_{||} = (k_x, k_y)$ a two–component vector in the x–y–plane.

One of the most sophisticated k*p Hamiltonians for calculation of the band structure can be derived by employing the 8 band–edge Bloch functions (see Table 2.1), which we introduced above. Then, the Schrödinger equation for an electron (or hole) in the bulk crystal is a set of 8 coupled differential equations, as we have already learned. If we now consider a heterostructure, more specifically a quantum well as shown in Fig. 2.8, we have to replace in the Hamiltonian k_z by $-i\partial/\partial z$, because of the quantization in z direction. Furthermore, we have to add the heterostructure (quantum well) potential $V_{\text{ext}}(z)$, which, both for the conduction band and the valence band, are square–well potentials as displayed schematically in Fig. 2.8(a). Hence, the full Schrödinger equation reads

$$\left[H_{8-\text{band}}(\boldsymbol{k}_{||}, -i\partial/\partial z) + V_{\text{ext}}(z)\right] \psi(\boldsymbol{r}) = E_l(\boldsymbol{k}_{||})\psi(\boldsymbol{r}) \; . \tag{2.35}$$

For a complete description of the quantum–well problem, the z–dependent parts of the envelope functions[7], $\bar{\chi}_l^{(A,B)}(z)$, of the materials A and B have to satisfy boundary conditions at the interfaces ($z = z_0$):

$$\bar{\chi}_l^{(A)}(z_0) = \bar{\chi}_l^{(B)}(z_0) \qquad \text{(continuity)} \tag{2.36}$$

$$\hat{A}^{(A)}\bar{\chi}_l^{(A)}(z_0) = \hat{A}^{(B)}\bar{\chi}_l^{(B)}(z_0)\,, \tag{2.37}$$

where the $\hat{A}^{(A,B)}$ are 8×8 differential operators. This 8–band model is known as the so called *Pidgeon–Brown model* [21].

There are a couple of useful approximations to this relatively complicated problem. In many practical cases, if electrons in the conduction band are considered, only, it is sufficient to use the parabolic approximation for the conduction band dispersion in the vicinity of the Γ–point, and neglect coupling to other bands. In this case, which is known as the *Ben Daniel–Duke model* [22], after separation of the inplane motion, the Schrödinger equation (2.35) simplifies drastically to two independent differential equations, one for each spin direction of the electron

$$\left[-\frac{\hbar^2}{2}\frac{\partial}{\partial z}\frac{1}{\mu(z)}\frac{\partial}{\partial z} + V_{\text{ext}}(z) + \frac{\hbar^2 k_{||}^2}{2\mu(z)}\right]\chi_{\pm 1/2,n}(z) = E_n(\boldsymbol{k}_{||})\chi_{\pm 1/2,n}(z)\,. \tag{2.38}$$

Here, $\mu(z)$ is the z–dependent effective mass, $\mu(z) = m_A^*$ ($\mu(z) = m_B^*$), if the electron is in material A (B). In the Ben Daniel–Duke model, the boundary conditions (2.36) and (2.37) simplify to

$$\chi_{\pm 1/2,n}^{(A)}(z_0) = \chi_{\pm 1/2,n}^{(B)}(z_0) \qquad \text{(continuity)} \tag{2.39}$$

$$\frac{1}{m_A^*}\frac{\partial}{\partial z}\chi_{\pm 1/2,n}^{(A)}(z_0) = \frac{1}{m_B^*}\frac{\partial}{\partial z}\chi_{\pm 1/2,n}^{(B)}(z_0)\,, \tag{2.40}$$

with scalar functions for the $\chi_{\pm 1/2,n}^{(A,B)}$. These boundary conditions are exactly the same as can be found in standard quantum mechanics textbooks for the problem of an electron in a one–dimensional square–well potential, except for the prefactors, which take care of the different masses in materials A and B.

For many cases it is also sufficient to consider for the valence band the 4 Γ_8 Bloch functions (for $J_z = \pm\frac{3}{2}$ and $J_z = \pm\frac{1}{2}$), only, and neglect coupling to the conduction band and to the split–off valence band. In this case, the boundary conditions for holes are such that the following components of the $\chi_l(z)$ and of their derivatives have to satisfy continuity conditions at the interfaces, $z = z_0$, between materials A and B [12]

$$\begin{bmatrix} i(\gamma_1 - 2\gamma_2)\frac{\partial}{\partial z} & \sqrt{3}\gamma_3(k_x - ik_y) \\ \sqrt{3}\gamma_3(k_x + ik_y) & i(\gamma_1 + 2\gamma_2)\frac{\partial}{\partial z} \end{bmatrix} \begin{pmatrix} \chi_{3/2}(z) \\ \chi_{1/2}(z) \end{pmatrix}, \tag{2.41}$$

[7] In an 8–band theory, the $\bar{\chi}_l(z)$ are eight–component spinors.

$$\begin{bmatrix} i(\gamma_1 + 2\gamma_2)\frac{\partial}{\partial z} & -\sqrt{3}\gamma_3(k_x - ik_y) \\ -\sqrt{3}\gamma_3(k_x + ik_y) & i(\gamma_1 - 2\gamma_2)\frac{\partial}{\partial z} \end{bmatrix} \begin{pmatrix} \chi_{-1/2}(z) \\ \chi_{-3/2}(z) \end{pmatrix} . \tag{2.42}$$

So far, we have considered only a single electron (hole) in the conduction (valence) band and neglected the interaction of this charge carrier with all the other electrons, or with the ionized donors in the barrier layers. We will take care of these interaction effects on the ground state of a semiconductor heterostructure in the following subsection.

2.4.2 Self–Consistent Band Structure Calculation

In this subsection we will outline briefly the most commonly used methods to calculate the ground state of a doped semiconductor heterostructure, taking into account Coulomb interaction of the charge carriers. We will restrict ourselves here to the consideration of electrons in the conduction band, with effective mass m^*, only. In principle, similar calculations for holes are straight forward, using the more involved boundary conditions, which were introduced above. To solve the problem exactly, one would have to deal with the Hamiltonian of the many–particle system, including Coulomb interaction[8]

$$H = \sum_i \left\{ -\frac{\hbar^2}{2m^*}\Delta_i + V_{\text{ext}}(\boldsymbol{r_i}) \right\} + \frac{e^2}{4\pi\epsilon\epsilon_0}\frac{1}{2}\sum_{i\neq j}\frac{1}{|\boldsymbol{r_i} - \boldsymbol{r_j}|} , \tag{2.43}$$

where i and j run over all charge carriers in the system. This problem, however, is exactly solvable for small electron numbers N, typically $N < 10$, only. For such exact solutions of the full Hamiltonian, very often the technique of numerical diagonalization is employed (see, e.g., [23, 24]).

For most of the theoretical calculations for larger systems, so called mean–field approaches are applied. In a mean–field approach, the many–particle problem is reduced to an *effective* single–particle one. The idea behind this is that the electron moves in a potential which consists of the external (quantum well) potential plus a potential, which is formed by all the other electrons and the ionized impurities. This means that one has to solve an effective one–electron Schrödinger equation and the Poisson equation of the structure self–consistently. The simplest form, where only the direct, i.e., classical, Coulomb interaction is taken into account, is called the Hartree approximation (HA). The effective one–electron Schrödinger equation in the HA reads

$$\left\{ -\frac{\hbar^2}{2m^*}\Delta + V_{\text{ext}}(\boldsymbol{r}) + \frac{e^2}{4\pi\epsilon\epsilon_0}\sum_\nu f_\nu \int \frac{|\psi_\nu(\boldsymbol{r}')|^2}{|\boldsymbol{r} - \boldsymbol{r}'|} d^3\boldsymbol{r}' \right\} \psi_n(\boldsymbol{r}) = E_n\psi_n(\boldsymbol{r}) , \tag{2.44}$$

[8] For simplicity, we have omitted here the explicit inclusion of spin–orbit interaction.

with the last term on the left–hand side being the Hartree potential, $V_{\mathrm{H}}(\boldsymbol{r})$. The quantum numbers ν are combined indices (e.g., spin and subband quantum numbers), and f_ν is the Fermi occupation number of the state $|\nu\rangle$. There are several standard methods to include quantum mechanical corrections to the Coulomb interaction. The straight forward method is the so called Hartree–Fock approximation (HFA), where, additionally, an exchange term is included. However, such calculations are relatively involved, since the exchange – or Fock – term is nonlocal, as can be seen by comparing the following Schrödinger equation in the HFA,

$$\left\{-\frac{\hbar^2}{2m^*}\Delta + V_{\mathrm{ext}}(\boldsymbol{r}) + \frac{e^2}{4\pi\epsilon\epsilon_0}\sum_\nu f_\nu \int \frac{|\psi_\nu(\boldsymbol{r}')|^2}{|\boldsymbol{r}-\boldsymbol{r}'|}d^3\boldsymbol{r}'\right\}\psi_n(\boldsymbol{r}) - \\ -\frac{e^2}{4\pi\epsilon\epsilon_0}\sum_\nu f_\nu \int \frac{\psi_\nu^*(\boldsymbol{r}')\psi_n(\boldsymbol{r}')}{|\boldsymbol{r}-\boldsymbol{r}'|}d^3\boldsymbol{r}'\psi_\nu(\boldsymbol{r}) = E_n\psi_n(\boldsymbol{r})\,, \quad (2.45)$$

with (2.44). In a systematic perturbation expansion with respect to Coulomb interaction, the Fock term is representing the first quantum mechanical correction – called *exchange interaction* – to the classical problem of an interacting many–particle system. The reason for this correction is the indistinguishability of quantum–mechanical particles. The not exactly known infinite sum of all additional corrections is called *correlations*. We will not further discuss the HFA approach in this introductory chapter.

A frequently used simpler method is to include exchange plus correlation corrections by adding a local potential which can be derived from the local–density approximation (LDA) of the density–functional theory of Hohenberg, Kohn, and Sham [25, 26, 27]. This so called Kohn–Sham approximation has the advantage that the additional potential, $V_{\mathrm{XC}}(\boldsymbol{r})$, is local in $\boldsymbol{r}$, and hence can be treated just as an additive contribution to the Hartree potential $V_{\mathrm{H}}(\boldsymbol{r})$. Within this model, the many–particle Schödinger equation (2.43) reduces to the so called *Kohn–Sham equation*, an equation for a single electron moving in an effective potential, $V_{\mathrm{LDA}}(\boldsymbol{r}) = V_{\mathrm{H}}(\boldsymbol{r}) + V_{\mathrm{XC}}(\boldsymbol{r})$,

$$\left\{-\frac{\hbar^2}{2m^*}\Delta + V_{\mathrm{LDA}}(\boldsymbol{r}) + V_{\mathrm{ext}}(\boldsymbol{r})\right\}\psi_n(\boldsymbol{r}) = E_n\,\psi_n(\boldsymbol{r})\,. \quad (2.46)$$

In the LDA, the exchange–correlation energy of a homogeneous electron system is used. $V_{\mathrm{XC}}(\boldsymbol{r})$ depends on the local density $n(\boldsymbol{r})$ and the local spin density $\zeta(\boldsymbol{r})$. It is usually taken from quantum Monte Carlo calculations [28].

Applying the Ben Daniel–Duke boundary conditions (2.39) and (2.40) to (2.46), and separating the free electron motion parallel to the quantum well plane, leads to the one–dimensional Schrödinger equation

$$\left\{-\frac{\hbar^2}{2}\frac{d}{dz}\frac{1}{m^*(z)}\frac{d}{dz} + \frac{\hbar^2k_{||}^2}{2m^*(z)} + V_{\mathrm{LDA}}(z) + V_{\mathrm{ext}}(z)\right\}\chi_n(z) = E_n\,\chi_n(z)\,, \quad (2.47)$$

with

$$V_{\mathrm{LDA}}(z) = V_{\mathrm{H}}(z) + V_{\mathrm{XC}}(z) \,. \tag{2.48}$$

The Hartree energy is given by the electrostatic potential $\phi(z)$, which acts on the electron

$$V_{\mathrm{H}}(z) = -e\phi(z) \,, \tag{2.49}$$

which, on the other hand, is the solution of the Poisson equation

$$\epsilon_0\epsilon \frac{d^2\phi(z)}{dz^2} = e\,n(z) - \rho(z) \,. \tag{2.50}$$

In (2.50), $\rho(z)$ is the charge distribution of ionized donors and acceptors, and,

$$n(z) = \sum_i N_i \left|\chi_i(z)\right|^2 \,, \tag{2.51}$$

with N_i being the number of electrons in the i–th subband. From (2.49)–(2.51) it follows that the Hartree potential $V_{\mathrm{H}}(z)$, which enters the Kohn–Sham equation (2.47) via (2.48), itself depends on the solutions, $\chi_i(z)$, of (2.47). Therefore, an iterative *self-consistent* solution of both, the Kohn–Sham and the Poisson equations has to be performed.

Figure 2.9 displays the results of such a calculation for a one–sided modulation–doped GaAs–AlGaAs single quantum well, as shown in Fig. 2.8.

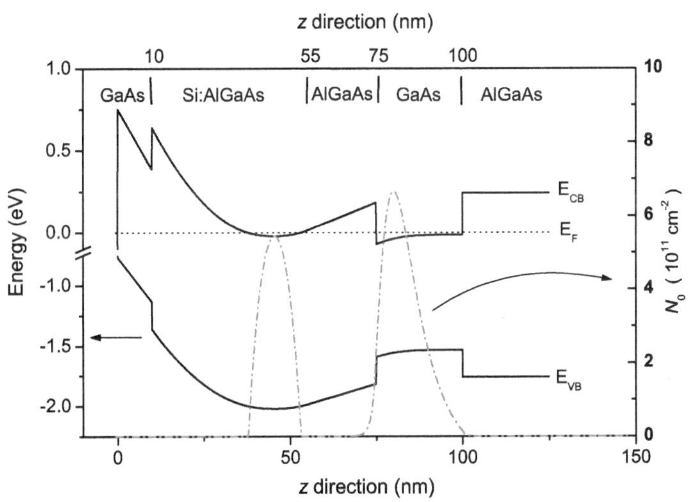

Fig. 2.9. Self–consistent Kohn–Sham calculation of the valence– and conduction–band edges of a one–sided doped GaAs–AlGaAs single quantum well as typically used for inelastic light scattering experiments (calculation after [29]). The *dashed–dotted lines* indicate the densities of free electrons. In the quantum well, as well as in the doped barrier layer, there is a finite density of free electrons in the conduction band

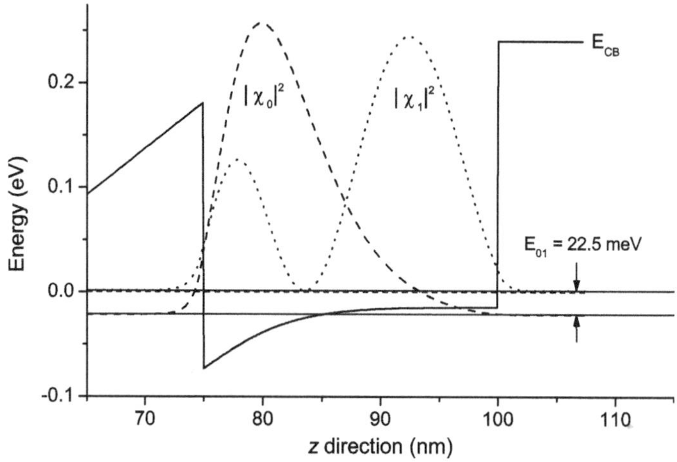

Fig. 2.10. Self-consistently calculated conduction band profile of a GaAs–AlGaAs single quantum well. The two lowest confined levels are indicated. The *dashed* and *dotted lines* represent the squares of the envelope wave functions $|\chi_0|^2$ and $|\chi_1|^2$, respectively

Such asymmetric quantum wells are frequently used samples for inelastic light scattering experiments. In Fig. 2.9, the valence– and conduction–band edges of the sample are shown. The dashed–dotted lines indicate the densities of free electrons, which result from the modulation doping. Figure 2.10 shows the well in the conduction band in more detail. The two lowest confined subband levels are indicated together with the squared envelope wave functions $|\chi_0|^2$ and $|\chi_1|^2$, which give the probability densities of the electrons in the states.

In Fig. 2.11, the subband spacing, $E_1 - E_0$, between the first two subbands of the quantum well in the conduction band is plotted versus the total density, $n = N_0 + N_1$, of electrons of the 2DES in the quantum well. The full line shows the results of a Kohn–Sham calculation, and the dashed line gives the results derived in Hartree approximation, i.e., neglecting exchange–correlation corrections. One can see that with increasing total electron density n, the subband spacing $E_1 - E_0$ becomes larger. This is due to the stronger band bending in the quantum well with increasing n (cf. Fig. 2.10). At the densities where the kinks in the calculated curves appear – which is in the Hartree calculation at $n \sim 6 \times 10^{11}$ cm^{-2} and in the Kohn–Sham calculation at $n \sim 8 \times 10^{11}$ cm^{-2} – the second subband, E_1, starts to be filled with electrons. This causes a redistribution of the carriers in the quantum well and thus influences the self–consistent potential.

The dispersion of the electronic subbands with respect to the inplane quasi momentum $\boldsymbol{k}_{\|} = (k_x, k_y)$ is to a good approximation parabolic, if the coupling to the valence–band states is neglected. For the confined levels in the

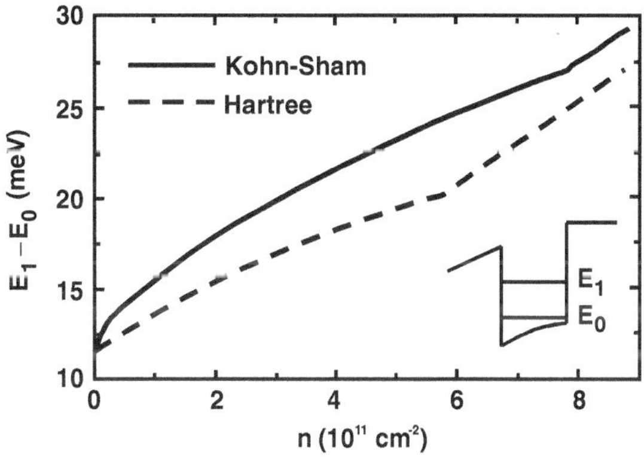

Fig. 2.11. Comparison of self–consistent calculations in the Hartree (dashed line) and LDA approximation (*solid line*) of the subband separation, $E_1 - E_0$, of a 25 nm wide GaAs–AlGaAs single quantum well in dependence on the total density of Q2D electrons, $n = N_0 + N_1$, in the two lowest subbands (data after [30])

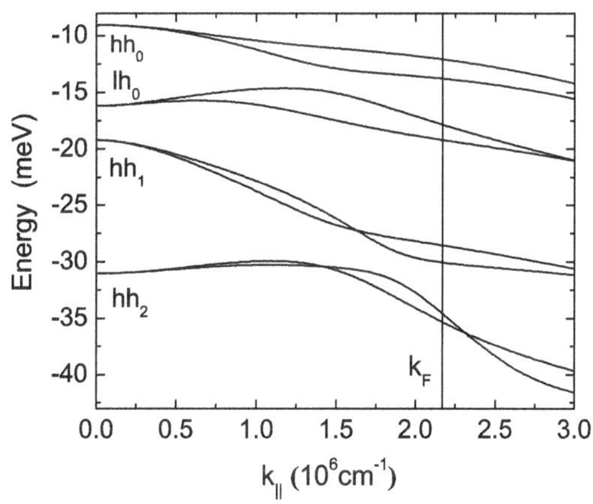

Fig. 2.12. Calculated wave–vector dispersion of hole subbands in an asymmetric valence–band single quantum well, as, e.g., displayed in Fig. 2.9. The levels are labeled – due to their surviving character at $\boldsymbol{k}_{||} = 0$ – as hh_i and lh_i for heavy– and light–hole states, respectively

valence–band well, however, the dispersion is strongly nonparabolic due to the above described effect of heavy– and light–hole mixing at $\boldsymbol{k}_{||} \neq 0$. Figure 2.12 displays the inplane wave–vector dispersion of the first four confined hole levels in a valence–band well, as shown in Fig. 2.9. The hole subbands are calculated employing a 4×4 Luttinger–Kohn Hamiltonian, as described in Sect. 2.1.2 above. For the system shown in Fig. 2.12, the self–consistent density of Q2D electrons was 7.5×10^{11} cm^{-2}. The position where the Fermi wave vector $\boldsymbol{k_F}$ of the electrons is, is indicated in the figure by a vertical line. The hole levels exhibit a spin splitting for finite $\boldsymbol{k}_{\|}$ due to the combined effect of spin–orbit coupling and the asymmetric confining potential (for more details, see, e.g., [3]). The hole states are important in the resonant inelastic light scattering process, since there they serve as intermediate states.

We will use a Q2D electron gas in a quantum well, as displayed in Figs. 2.9 to 2.11, later in Chap. 3 to introduce the electronic elementary excitations of a 2DES.

2.5 Lateral Micro- and Nanostructures

2.5.1 General Remarks

The traditional way of producing lateral semiconductor nanostructures, like quantum wires and quantum dots, is to pattern down MBE–grown Q2D systems. There is a large variety of possibilities to do that[9]. The principal processing steps, however, are quite similar in all the possible variations, and are also established in industry as the key processes for the production of devices. Figure 2.13 shows an overview of the most important steps in lateral patterning. Starting from the MBE wafer, which, e.g., can be a modulation–doped GaAs–AlGaAs structure, in the first step, a resist layer is deposited. This resist can either be a photoresist, or an electron–beam sensitive resist, like polymethylmethracrylate (PMMA). Subsequently, the resist–coated sample is exposed to a beam pattern of light, or to an electron–beam. After exposure, a resist pattern is obtained in a development process. Now, there are several further ways to proceed with such a resist pattern. One possibility is to use it as an etch mask to transfer the pattern into the semiconductor surface by an etching process. An other possibility is to invert the resist mask into a patterned metal film, by depositing a metal (e.g., Al, Ti, or NiCr), and subsequently removing those parts of the metal film, which are deposited on top of the resist by an etching process, which removes the resist. This procedure is called a *lift–off process.* The patterned metal film may then either be used again as a mask, e.g., for ion implantation, or as a gate electrode. In the case of patterning the semiconductor surface by etching, there are two established procedures. One is, to etch completely through the Q2D electron system. This processing is known as *deep etching.* Another possibility, which is called *shallow etching,* is to stop the etching process in the layer, which contains the dopants (e.g., Si in an AlGaAs barrier layer). In the case of shallow etching, the lateral structure is defined by the periodic electrostatic potential, given by the ionized donors, only. In Fig. 2.14, the processing steps, which are involved in a patterning by etching, are schematically displayed in

[9] For a larger overview over these processes, see, e.g., the more general monographs, [31, 32].

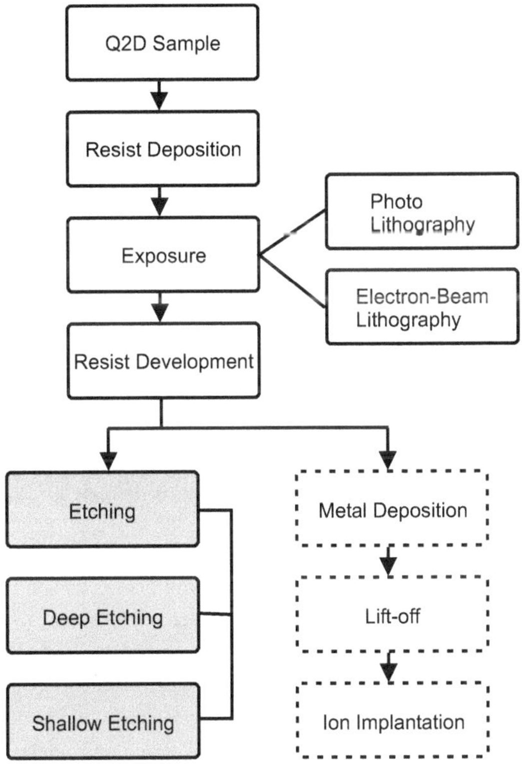

Fig. 2.13. Block diagram of the most important steps in the fabrication of lateral micro– and nanostructures

more detail. Those are the key processes, which were used for the preparation of deep–etched quantum–wire and quantum–dot samples, which will be considered later in this book in Chaps. 5, 6 and 7.

We will concentrate in the following on a laser–interference lithography process, which enables one to produce large periodic arrays – in the range of millimeters squared – of quantum wires or quantum dots. Such large arrays of nearly identical quantum structures are in particular well–suited for optical experiments, where one needs large active sample areas. In the next subsection, we will describe this optical lithography process, as well as the typically–used etching processes, in more detail.

2.5.2 Lithography and Etching

Laser–Interference Lithography

A very convenient way to produce large periodic arrays of nanostructures is to apply the interference pattern of two coherent laser beams. This method is

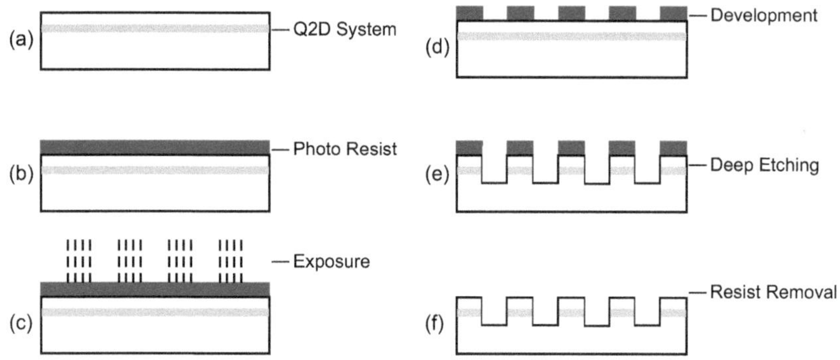

Fig. 2.14. Schematic representation of the processing steps for the preparation of etched quantum wires or quantum dots

called *holographic* or *interferometric lithography*. Historically, it was invented at the end of the 1960's for the fabrication of optical gratings. Later, the method was also applied for the definition of periodic photo–resist patterns on semiconductor surfaces, as considered in this book, which are subsequently used as etch masks for further processing. The etch–mask technique was applied in the 1980's for the preparation of modulated 2D structures (e.g., [33]), semiconductor quantum wires (e.g., [34]), and quantum dots (e.g., [35, 36]). With shadowing techniques it is also possible to deposit a metal stripe array, which, e.g., can be used as a grating coupler (cf. Sect. 4.10) for the excitation of 2D plasmons or intersubband resonances in far–infrared absorption experiments (e.g., [37]).

Figure 2.15 displays schematically a setup, which can be used for laser–interference lithography. A photograph of a real setup is shown in Fig. 2.16. The beam of a laser is expanded to a diameter of several centimeters and split into two beams by a large–area beam splitter. Via two mirrors, the beams

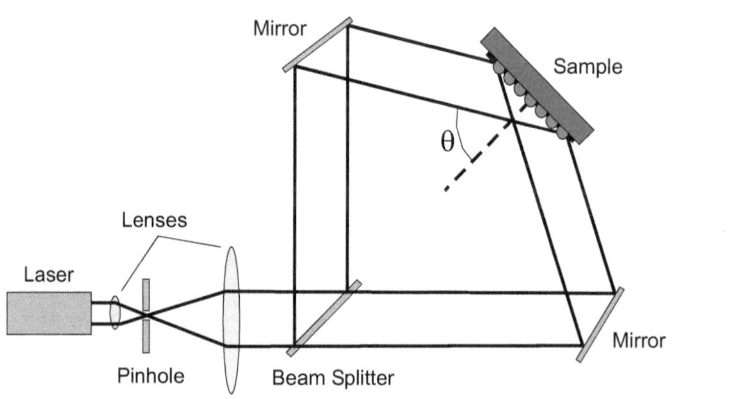

Fig. 2.15. Schematic picture of a setup for laser–interference lithography

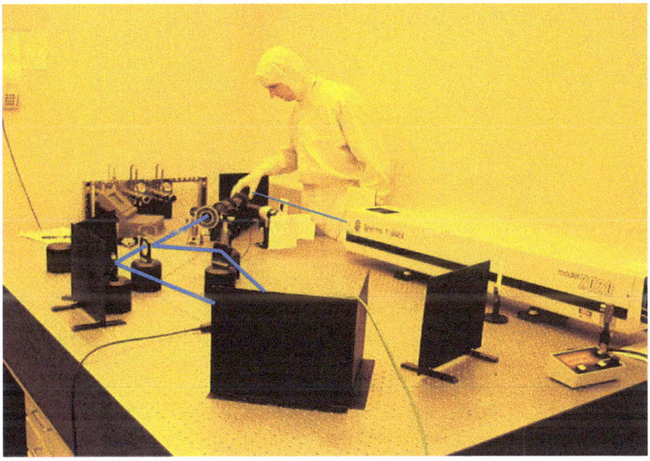

Fig. 2.16. Photograph of the laser–interference–lithography setup in the cleanroom of the Institute of Applied Physics at the University of Hamburg

are reflected onto the sample surface, which is coated with a photoresist. Due to constructive and destructive interference of the two laser beams, there are stripes with high and low intensity created at the photoresist. The period, a, of the stripe pattern is given by the laser wavelength, λ, and the angle, θ, (see Fig. 2.15) by the relation

$$a = \frac{\lambda}{2\eta} \sin\theta \,, \tag{2.52}$$

where η is the refractive index of the surrounding medium. With, e.g., the 364 nm UV line of an Argon laser, the minimum period length would be $a = 182$ nm. By variation of the exposure and development times, typically structure sizes of down to about 100 nm can be reproducibly achieved. Patterns for the preparation of dots can be realized by two exposure processes: After the first exposure, the sample is rotated by 90 degrees, and a second exposure is performed. By choosing the exposure and development times, either a dot–like or an antidot–like pattern can be produced. Here, an array of holes in the resist is called an antidot–like pattern. Figure 2.17 shows examples of a dot–like [Fig. 2.17(a)] and an antidot–like resist pattern [Fig. 2.17(b)] after the development process.

Wet and Dry Etching

There is a variety of methods to transform the resist pattern into the sample surface. GaAs is, e.g., wet–chemically etched by a solution of H_2SO_4, H_2O_2, and H_2O. With this solution, the etching rate is about 10 nm per second, if the constituents have a ratio of 1:8:1000. For the preparation of sub–micron

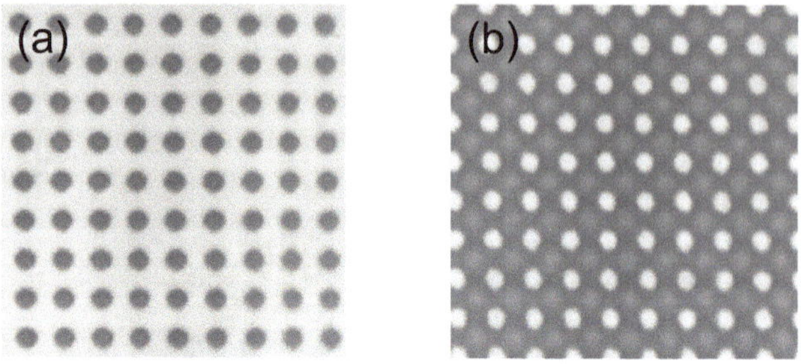

Fig. 2.17. Microscope images of resist patterns for (**a**) a dot array, and, (**b**) an antidot array, defined with laser–interference lithography. The photoresist appears dark gray [with courtesy of T. Kipp, Hamburg]

structures, this procedure is not perfectly suited because of the isotropy of the etching process. This isotropy results in an under etching of the mask, as illustrated in Fig. 2.18 (a). Beside wet–chemical etching, there is also the possibility to use gaseous radicals in a dry–etching plasma process. As etching gases, O_2, F_2, or Cl_2 plasmas are used.

Figure 2.18 (b) shows a schematic picture of a physical etching process, where, e.g., positively charged Ar^+ ions are accelerated onto the sample. Such processes are highly anisotropic but also non selective. The mask is etched with essentially the same rate as the semiconductor. Furthermore, it can produce a large density of defects in the material.

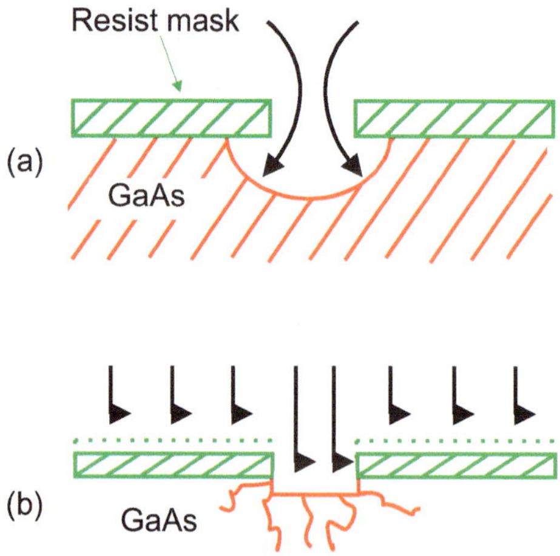

Fig. 2.18. Schematic pictures of (**a**) a wet–chemical, and, (**b**) a physical etching process through a resist mask

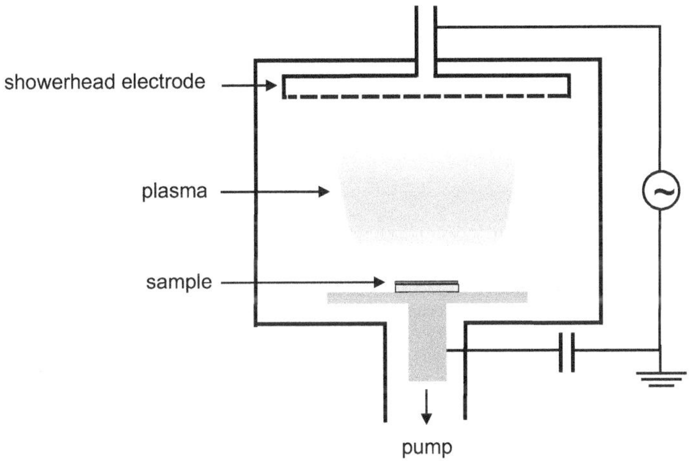

Fig. 2.19. Schematic picture of a reactive–ion etching reactor

Reactive–Ion Etching (RIE)

To get a highly anisotropic process with a minimum damage of the crystal, it is advantageous to combine chemical and physical etching processes. This is done in the reactive–ion etching (RIE) process. Figure 2.19 shows a schematic picture of a RIE reactor. Between top and bottom electrodes, a high–frequency (∼MHz) ac voltage is applied. The high–frequency voltage causes an ionization of the gas (e.g., $SiCl_4$) inside the reactor and creates free radicals and electrons. The light electrons can easily follow the ac voltage and lead to a charging of the electrodes. The high voltage is, however, coupled via a capacitor to the bottom electrode, i.e., the charge is kept on this electrode. This results in the built up of a dc voltage between plasma and bottom electrode. The heavy ions are not able to follow the high frequency and are accelerated by the dc voltage towards the bottom electrode, where the sample is mounted. Typically, the parameters are chosen such that most of the etching is achieved by chemical etching via the radicals with a minimal creation of defects. The RIE process has a quite high selectivity of about 10:1 in etching of the semiconductor as compared to the etching of the resist.

Figure 2.20 shows scanning electron micrographs of quantum–dot and of a quantum–wire array, which were prepared by laser–interference lithography and a RIE etching process.

2.5.3 Self–Assembled Quantum Dots

Since the end of the 1990's, an alternative method for the creation of quantum dots has been established and become very prominent: The *self–organized* or *self–assembled* growth[10]. The most prominent examples are InAs or InGaAs

[10] For an overview, see the monograph, [38].

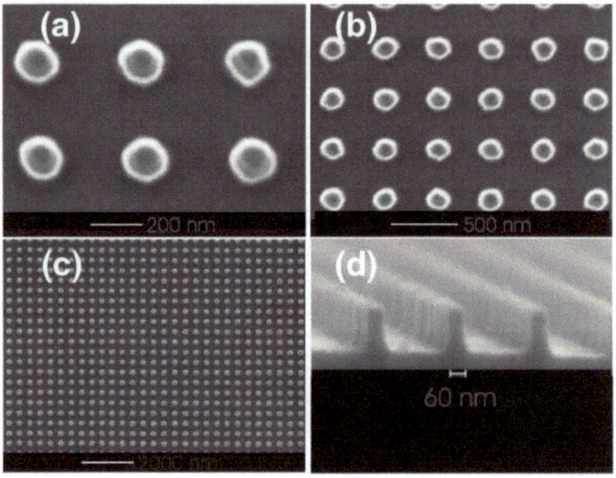

Fig. 2.20. Scanning electron micrographs of deep–etched GaAs–AlGaAs nanostructures

self–assembled quantum dots (SAQD), grown on GaAs. InAs and GaAs have strongly different lattice parameters (see Fig. 2.7). In the Stranski–Krastanow growth mode, InAs grows on GaAs pseudomorphically up to a critical thickness of about 1.5 monolayers. Above this critical thickness, it is energetically more favorable for the InAs to form small islands in order to reduce the stress. Figure 2.21 shows an atomic–force microscope image of MBE–grown InAs SAQD on a GaAs surface. Typically, the SAQD have lateral dimensions of about 20 nm and a height of about 7 nm. In Fig. 2.22, a transmission electron microscope cross section of a single dot is shown. Of course, there

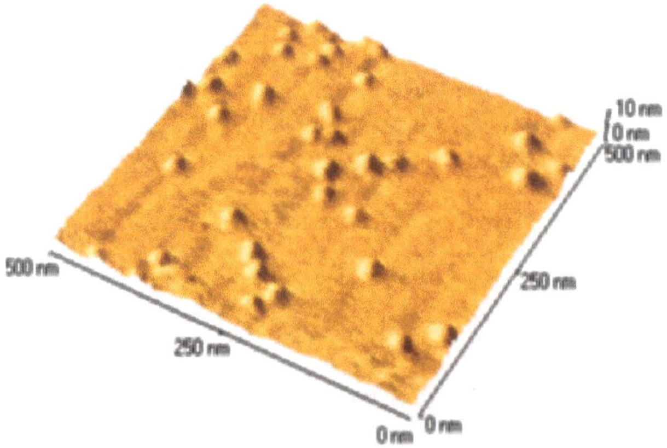

Fig. 2.21. Atomic–force microscope image of InAs quantum dots on GaAs

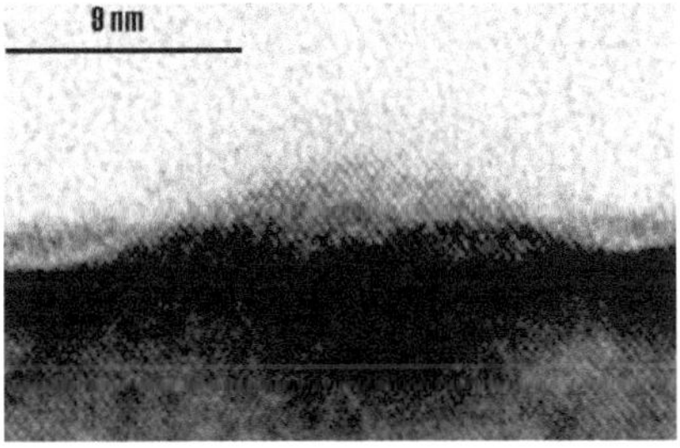

Fig. 2.22. Transmission electron micrograph of the cross section of a single InGaAs quantum dot [with permission of S. Mendach, Hamburg]

is a variety of possible sizes and shapes of these dots, depending on the detailed growth conditions. Nowadays, the growth with relatively narrow size distributions is standard. Over the years it was found that, depending on the detailed growth parameters, the SAQD can have different shapes, ranging from pyramidal to lens–shaped structures. InAs SAQD have become the prototype structures for the investigation of Q0D systems by optical means. During the past years, the photoluminescence spectroscopy of single SAQD has been well established and has become a standard investigation tool[11]. In inelastic light scattering experiments, the intensities are, however, usually several orders of magnitude smaller than in photoluminescence. Therefore, there are so far no single–dot inelastic light scattering experiments available. Experiments on ensembles of SAQD shall be discussed later in this book in Chap. 5.

2.6 Electronic Ground State of Lateral Nanostructures

The quantization energies due to the vertical confinement are for most of the lithographically defined structures much larger than the typical 1D subband spacings in quantum wires or level spacings in quantum dots. Therefore, in most of the theoretical treatments, quantum wires and quantum dots are considered to be strictly flat in z direction (growth direction of the vertical structure). In this case, the methods as introduced in Sect. 2.4 for Q2D systems for the self–consistent band structure calculation of the electronic states in the conduction band can be straight forwardly applied to Q1D or Q0D structures. The z direction can be separated out, and the EFA can be applied to the lateral directions. Then, the ground state can be calculated either employing the HA, HFA, or a Kohn–Sham–type calculation, as described in Sect. 2.4. The effect of the finite width of the nanostructures in *vertical* direction is discussed in [40]. For quantum dots with only few electrons, the

[11] For a recent collection of overview articles, see [39].

ground state can also being calculated by the so called exact numerical diagonalization of the few–electron Hamiltonian (see, e.g., [41, 42, 43, 44, 45, 46]). At the moment, such calculations are feasible up to electron numbers of about 10, only. In the chapter about inelastic light scattering experiments on quantum dots (Chap. 5) we will come back to this method. More details about the structure of the electronic ground state in quantum wires and quantum dots will be discussed directly in the following respective chapters about experiments on those systems. We note here that for the more involved calculation of the valence– and conduction–band states in not intentionally doped SAQD's, also 8 band $\boldsymbol{k}\cdot\boldsymbol{p}$ calculations have been applied [47].

References

1. Peter Y. Yu, Manuel Cardona: *Fundamentals of Semiconductors*, 1st edn (Springer, Berlin Heidelberg 1996) pp. 26ff
2. G. Dresselhaus: Phys. Rev. **100**, 580 (1955)
3. R. Winkler: *Spin-Orbit Coupling Effects in Two-Dimensional Electron and Hole Systems*, Springer Tracts in Modern Physics, Vol. 191, 2003
4. M. L. Cohen, J. Chelikowsky: *Electronic Structure and Opical Properties of Semiconductors*, 3nd edn., Springer Ser. Solid–State Sci., Vol. 75 (Springer, Berlin, Heidelberg 1989) p. 103
5. J. M. Luttinger, W. Kohn: Phys. Rev. **97**, 869 (1955)
6. J. M. Luttinger: Phys. Rev. **102**, 1030 (1956)
7. E. O. Kane: J. Phys. Chem. Solids **1**, 249 (1957)
8. E. O. Kane: *Semiconductors and Semimetals*, Vol. 1, ed. R. K. Willardson and A. C. Beer (Academic Press, New York 1966) pp. 75
9. E. O. Kane: *Handbook on Semiconductors*, Vol. 1, ed. T. S. Moss (North Holland, Amsterdam 1982) pp. 193
10. G. Bastard: *Wave mechanics applied to semiconductor heterostructures* (Halsted press, John Wiley and Sons New York Chichester Brisbane Toronto Sigapore 1988) pp. 31ff
11. T. Ruf, M. Cardona: Phys. Rev. B **41**, 10747 (1990)
12. M. Altarelli, U. Ekenberg, A. Fasolino: Phys. Rev. B **32**, 5138 (1985)
13. M. J. Kelly: *Low-Dimensional Semiconductors, Materials, Physics, technology, Devices*, (Oxford University Press, New York 1995) pp. 27–33
14. Peter Y. Yu, Manuel Cardona: *Fundamentals of Semiconductors*, 1st edn (Springer, Berlin Heidelberg 1996) pp. 6–10
15. J. Menendez, A. Pinczuk, D. J. Werder, A. C. Gossard, J. H. English: Phys. Rev. B **33**, 8862 (1986)
16. N. T. Linh: Festkörperprobleme **23**, 227 (1983)
17. K. Hirakawa, Y. Hashimoto, T. Ikoma: Appl. Phys. Lett. **57** (24), 2555 (1990)
18. R. Dingle, H. L. Störmer, A. C. Gossard, and W. Wiegmann: Appl. Phys. Lett. **33**, 665 (1978)
19. T. Ando, A. B. Fowler, and F. Stern: Rev. Mod. Phys. **54**, 437 (1982)
20. G. Bastard: Phys. Rev. B **24**, 5693 (1981)
21. C. R. Pidgeon and R. N. Brown: Phys. Rev. **146**, 575 (1966)
22. D. J. Ben–Daniel and C. B. Duke: Phys. Rev. **152**, 683 (1966)

23. T. Chakraborty, K. Niemelä, and P. Pietiläinen: Phys. Rev. Lett. **78**, 4829 (1997)
24. Christoph Steinebach, Christian Schüller, and Detelf Heitmann: Phys. Rev. B **61**, 15600 (2000)
25. P. Hohenberg and W. Kohn: Phys. Rev. **136**, B 864 (1964)
26. W. Kohn and L. J. Sham: Phys. Rev. **140**, A 1133 (1965)
27. L. J. Sham and W. Kohn: Phys. Rev. **145**, 561 (1966)
28. B. Tanatar and D. M. Ceperley: Phys. Rev. B **39**, 5005 (1989)
29. I.-H. Tan, G. Snider, and E. Hu: J. Appl. Phys. **68**, 4071 (1990)
30. Christoph Steinebach: *Selbstkonsistente Berechnung des Potentialverlaufs und der optischen Anregungen in GaAs/AlGaAs–Heterostrukturen* Diplomarbeit, Universität Hamburg 1996, p. 13
31. M. J. Kelly: *Low-Dimensional Semiconductors, Materials, Physics, technology, Devices*, (Oxford University Press, New York 1995) pp. 61–74
32. C. Weisbuch, B. Vinter: *Quantum semiconductor structures, Fundamentals and Applications*, (Academic Press, San Diego 1995) pp.
33. U. Mackens, D. Heitmann, L. Prager, J. P. Kotthaus, and W. Beinvogl: Phys. Rev. Lett. **53**, 1485 (1984)
34. W. Hansen, T. P. Smith III, K. Y. Lee, J. A. Brum, C. M. Knoedler, J. M. Hong, and D. P. Kern: Phys. Rev. Lett. **62**, 2168 (1989)
35. C. Sikorski and U. Merkt: Phys. Rev. Lett. **62**, 2164 (1989)
36. B. Meurer, D. Heitmann, and K. Ploog, Phys. Rev. Lett: **68**, 1371 (1992)
37. D. Heitmann, J. P. Kotthaus, and E. G. Mohr: Solid State Commun. **44**, 715 (1982)
38. D. Bimberg, M. Grundmann, and L. Ledentsov: *Quantum Dot Heterostructures* (Wiley, New York, 1999)
39. P. Michler (Ed.): *S*ingle Dot Spectroscopy (Springer Tracts in Modern Physics, Berlin Heidelberg, 2004)
40. C. Steinebach, C. Schüller, G. Biese, D. Heitmann, and K. Eberl: Phys. Rev. B **57**, 1703 (1998)
41. P. A. Maksym and T. Chakraborty: Phys. Rev. Lett. **65**, 108 (1990)
42. M. Wagner, U. Merkt, and A. V. Chaplik: Phys. Rev. B **45**, 1951 (1992)
43. D. Pfannkuche, V. Gudmundsson, and P. A. Maksym: Phys. Rev. B **47**, 2244 (1993)
44. S.-R. Yang, A. H. MacDonald, and M. D. Johnson: Phys. Rev. Lett. **71**, 3194 (1993)
45. J. J. Palacios, L. Martin-Moreno, G. Chiappe, E. Louis, and C. Tejedor: Phys. Rev. B **50**, 5760 (1994)
46. P. A. Maksym and T. Chakraborty: Phys. Rev. B **45**, 1947 (1992)
47. O. Stier, M. Grundmann, and D. Bimberg: Phys. Rev. B **59**, 5688 (1999)

3 Electronic Elementary Excitations

In inelastic light scattering experiments on semiconductor nanostructures, electronic excitations are created or annihilated in the low–dimensional electron systems under investigation. Thus, the main body of this book will deal with the physics of those electronic elementary excitations in various systems and under various conditions. Before we elaborate on the basic concepts of the inelastic light scattering processes themselves in the following chapter, the electronic elementary excitations shall be introduced and discussed here. We will do this by the – most prominent – example of the excitations of Q2D electron systems, realized in modulation–doped GaAs–$Al_xGa_{1-x}As$ quantum wells. These excitations can be categorized into so called spin–density excitations (SDE) and charge–density excitations(CDE), which both are collective plasma oscillations of the Q2D system, and, single–particle excitations (SPE). In particular, the observation of intersubband SPE [1] – which are thought to be electronic excitations, which are not affected by the Coulomb interaction – has posed a puzzle, and has been controversially discussed. We will come to this discussion at various places later in this book, when considering the resonant scattering in quantum wells and in quantum dots. In particular in Chap. 5 we will see that – at least for quantum dots – the SPE's are actually *collective* excitations: SDE's and CDE's. However, the many–particle interaction effects partly cancel under specific conditions so that the energies are close to single–particle energies of a noninteracting system. Historically, in 1979, intersubband CDE and SDE in GaAs–AlGaAs quantum wells [2] and heterojunctions [3] were the first electronic excitations, which were observed in semiconductor nanostructures by inelastic light scattering by A. Pinczuk et al. and G. Abstreiter et al., respectively.

It shall be noted here that the excitation categories, which will be described in this section for Q2D systems, can be transferred to systems with lower dimensionality, i.e., quantum wires and quantum dots. The specialities of those systems will be introduced and discussed in the respective following chapters. Furthermore, we will in the current section already make use of polarization selection rules for the electronic excitations, and of the fact that in inelastic light scattering experiments a finite quasi momentum $\boldsymbol{q}$ can be transferred to the excitations. These characteristic features of the inelastic light scattering process itself will be introduced in detail later, in Chap. 4.

Christian Schüller: *Inelastic Light Scattering of Semiconductor Nanostructures*
STMP **219**, 41–55 (2006)
DOI 10.1007/3-540-36526-5_3

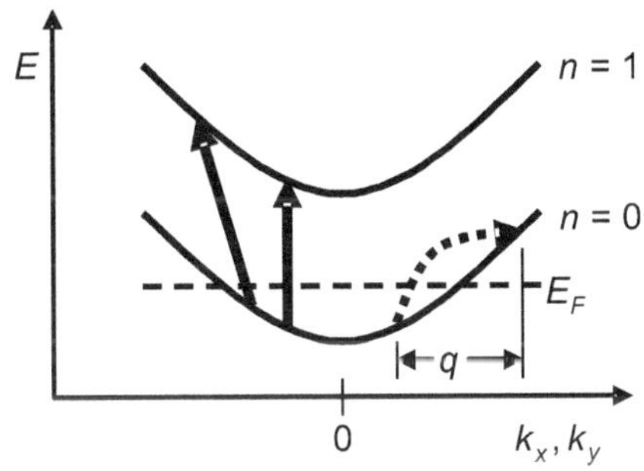

Fig. 3.1. Schematic picture of intersubband and intrasubband single–particle transitions in a Q2D electron system. The label k_x, k_y at the x axis shall indicate that for a Q2D system the dispersion of a subband is not a one–dimensional parabola but rather a paraboloid in k_x and k_y directions

3.1 Single–Particle Continua

We will start the discussion of elementary electronic excitations by considering a Q2D electron system with two subbands, where only the lowest subband is occupied by electrons up to the Fermi energy E_F at temperature $T = 0$ K, as displayed in Fig. 3.1. For the moment, the Coulomb interaction shall be neglected, and the electrons are considered to be independent particles. In this single–particle picture, we can think of *intersubband* and *intrasubband* SPE, as indicated in Fig. 3.1 by thick solid and dotted arrows, respectively. In the following – and for the rest of this book – we will always denote a quasimomentum or wave vector, which is connected to an electronic excitation, by $\boldsymbol{q}$, while the quasimomentum of an electron in the ground state will be labeled by $\boldsymbol{k}$ (for a Q2D system $\boldsymbol{k} \equiv \boldsymbol{k}_{\|} = (k_x, k_y)$). In Fig. 3.1 we can see that concerning intersubband excitations we can imagine transitions with finite wave–vector transfer q, and vertical transitions with $q = 0$. Intrasubband SPE, on the other hand, have nonzero energies for $q \neq 0$, only. All possible SPE, in dependence on the wave vector q, form the *single–particle continua.* Figure 3.2 displays schematically the intra– and intersubband single–particle continuum for the above introduced two subband system. The insets illustrate some selected situations at the edges of the continua. The upper edge of the intrasubband continuum is given by

$$E(k_F + q) - E(k_F) = \frac{\hbar^2 (k_F + q)^2}{2m^*} - \frac{\hbar^2 k_F^2}{2m^*} = \frac{q k_F}{m^*} + \frac{\hbar^2 q^2}{2m^*} \sim \frac{q k_F}{m^*} \, , \tag{3.1}$$

if $k_F \gg q$, i.e., the Fermi wave vector, k_F, is large compared to the wave vector, q, of the excitation. The lower bound of the continuum is given by zero for all $q < 2k_F$, since the 2D subband is a paraboloid, and so an intrasubband transition can start directly at the Fermi energy and end at an infinitesimally small energy just above the Fermi energy at a different point of the paraboloid

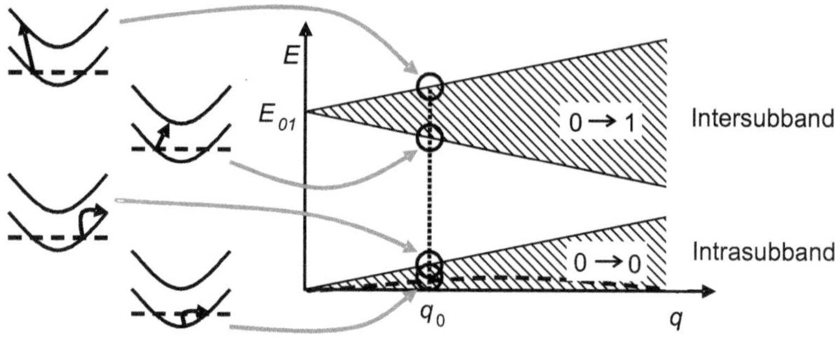

Fig. 3.2. Single–particle continua for intra– and intersubband transitions

in k_x and k_y (provided that $q < 2k_F$). For a Q1D electron system, however, which will be considered later in this book, the lowest subband is a one–dimensional parabola. In that case, the lower bound of the intrasubband continuum is given by the dashed line in Fig. 3.2, which is

$$E(k_F) - E(k_F - q) = \frac{qk_F}{m^*} - \frac{\hbar^2 q^2}{2m^*} \,. \tag{3.2}$$

For the upper and lower edges of the intersubband continuum, the energies are given by

$$E_{01} + E(k_F + q) - E(k_F) = E_{01} + \frac{qk_F}{m^*} + \frac{\hbar^2 q^2}{2m^*} \,, \tag{3.3}$$

and,

$$E_{01} - (E(k_F) - E(k_F - q)) = E_{01} - \frac{qk_F}{m^*} + \frac{\hbar^2 q^2}{2m^*} \,, \tag{3.4}$$

respectively. $E_{01} = E_1 - E_0$ is the subband spacing at any given, but fixed, inplane wave vector $\boldsymbol{k}_{\|}$.

Within this independent–particle model, one would thus expect for an experiment with a fixed finite wave–vector transfer q_0, the intra– and intersubband SPE's to exhibit a finite width, given by the widths of the respective continua, as , e.g., indicated in Fig. 3.2. For $q = 0$, we would expect a sharp intersubband SPE with energy E_{01}, and zero energy of the intrasubband SPE.

3.2 Electron–Density Waves: Phenomenology of Collective Charge– and Spin–Density Excitations

Figure 3.3 shows two inelastic light scattering spectra of intersubband excitations of a 25 nm–wide GaAs-AlGaAs single quantum well for a wave–vector transfer of $q \sim 0$. With the considerations about independent particles in

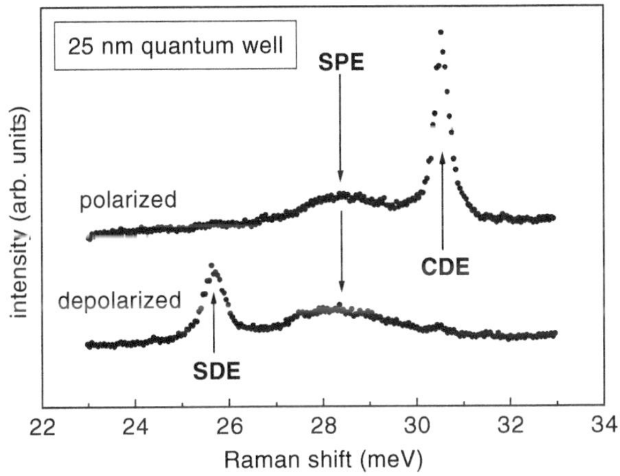

Fig. 3.3. Raman spectra of intersubband excitations in a quantum well. The 2D carrier density was $N_s = 7.5 \times 10^{11}$ cm^{-2}

the previous subsection, we would expect a sharp intersubband SPE at the single–particle subband spacing E_{01}. Obviously, the spectra in Fig. 3.3 are more complex. The spectra in Fig. 3.3 differ in the polarization directions of the incoming and the inelastically scattered light. In the *polarized* spectrum, the directions were parallel to each other, and, in the *depolarized* one, they are perpendicular. A broad feature is visible at approximately the same energetic position (28.3 meV) in both spectra. This feature appears at an energy, which one calculates for the mentioned bare subband spacing, E_{01}, of the two lowest subbands. Similar spectra of a high–mobility 2DES in a quantum well were observed for the first time in 1989 by A. Pinczuk et al. [1]. The broad feature is interpreted as an intersubband SPE. It is however surprising that, though we would expect a sharp line in an experiment at $q = 0$, the SPE has a considerable width. In addition, we observe in Fig. 3.3 two relatively sharp lines at energies above and below the SPE in the polarized and depolarized spectrum, respectively. Scattering geometries and selection rules will be explained in detail in the next chapter. We will, however, here already make use of the fact that, due to polarization selection rules, collective CDE should appear in polarized scattering geometry, and, collective SDE in depolarized geometry, only. Hence, we conclude that the many–particle intersubband excitations – the CDE and SDE – appear as narrow lines in the spectra.

CDE and SDE are collective plasma oscillations of the 2DES. A CDE – or plasmon – is a macroscopic oscillation of the charge density, which is induced by an external perturbation, like an electromagnetic dipolar field. The induced charge density is oscillating against the positively–charged background. In an intersubband CDE, as considered here, the sheet of Q2D

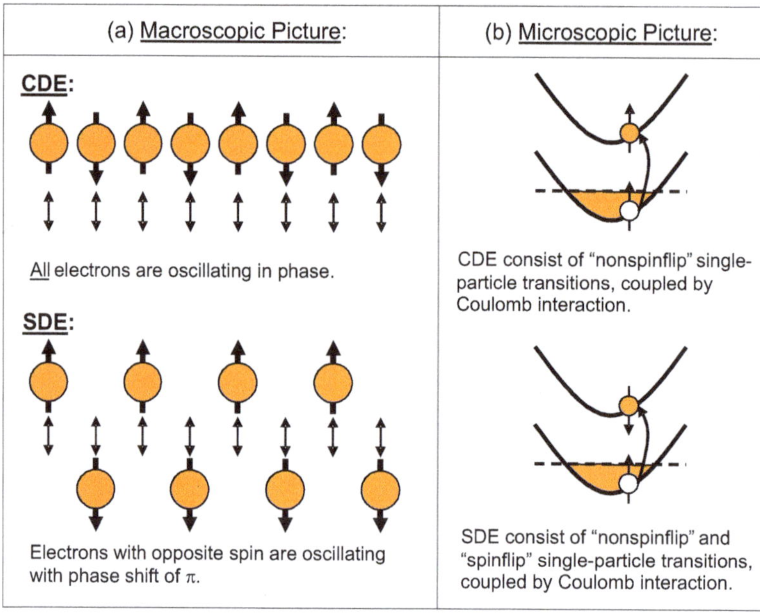

Fig. 3.4. Comparison of macroscopic and microscopic picture for collective intersubband excitations. In (**a**) the oscillation of the electron layer, parallel to the growth direction, is shown schematically

electrons oscillates *perpendicular* to the quantum–well plane. Thus, the center of mass of the charge is oscillating back and forth between the two interfaces of the quantum well. Figure 3.4(a) shows, very schematically, the oscillation of the electrons in this macroscopic picture. Microscopically, the intersubband CDE consists of non–spinflip intersubband transitions, as shown in Fig. 3.4(b), which are coupled by the full quantum–mechanical Coulomb interaction. Due to the depolarization field, caused by the direct part of the Coulomb interaction, the energy of the CDE is in most cases – so here – blueshifted with respect to the bare single–particle subband spacing. This will be explained in a more quantitative way in the following section. Due to its finite dipole moment, the intersubband CDE is far–infrared active, and can also be observed in far–infrared dipole–absorption experiments (see, e.g., [4]).

In the intersubband SDE – on the contrary – electrons with different spins are oscillating with a phase shift of π against each others (Fig. 3.4(a)). As a consequence, the center of mass of the charge does not oscillate. Hence, these excitations have a zero dipole moment, and are not subjected to depolarization–field effects. The SDE are only affected by exchange Coulomb interaction, which is attractive for electrons. Therefore, the energy of the SDE is redshifted with respect to the SPE. Microscopically, the intersubband SDE consists of intersubband spinflip and non–spinflip transitions, which are cou-

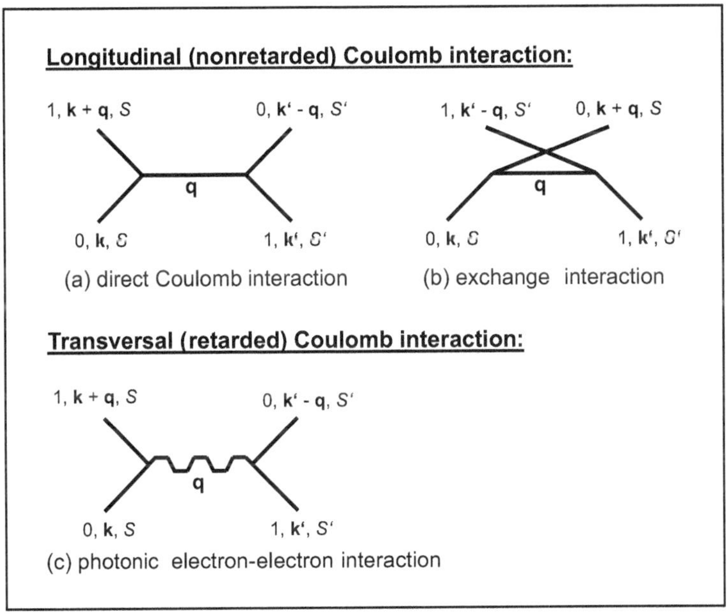

Fig. 3.5. Diagrammatic representations of Coulomb interaction for intersubband excitations (after [5]). The states are characterized by the quantum numbers $(n, \boldsymbol{k}, S)$, where n is the subband quantum number, $\boldsymbol{k}$ the inplane quasi momentum, and S the spin of the particle

pled by exchange interaction, only (see Fig. 3.4(b)). At zero magnetic field, the spinflip and non–spinflip single–particle transitions are energetically degenerate. Since the electric dipole operator does not act on the spin of the electrons, SDE can not be excited in direct absorption experiments. The creation of SDE is a speciality of the *resonant* inelastic light scattering process, as will be explained in detail in Chap. 4 below.

Figure 3.5 shows diagrams of the particle–particle interactions, which contribute to the formation of collective intersubband excitations of our two–subband system. The states, which participate in the interaction process, are characterized by the quantum numbers $(n, \boldsymbol{k}, S)$, where n is the subband quantum number, $\boldsymbol{k}$ the inplane quasimomentum, and S the spin of the particle. The direct Coulomb interaction (Fig. 3.5(a)), which leads to the formation of the CDE, one can imagine in the classical sense: By the interaction $(\boldsymbol{q})$, a particle is scattered by another particle from a state with subband label 0, momentum $\boldsymbol{k}$, and spin S to a state $(1, \boldsymbol{k} + \boldsymbol{q}, S)$, while the other particle is scattered from $(1, \boldsymbol{k}', S')$ to a state $(0, \boldsymbol{k}' - \boldsymbol{q}, S')$. If one would consider this part of the Coulomb interaction, only, one would make – quantum mechanically – an error, since, in quantum mechanics, particles are not distinguishable. From the principle of indistinguishability of particles, the exchange term (Fig. 3.5(b)) of the longitudinal Coulomb interaction results.

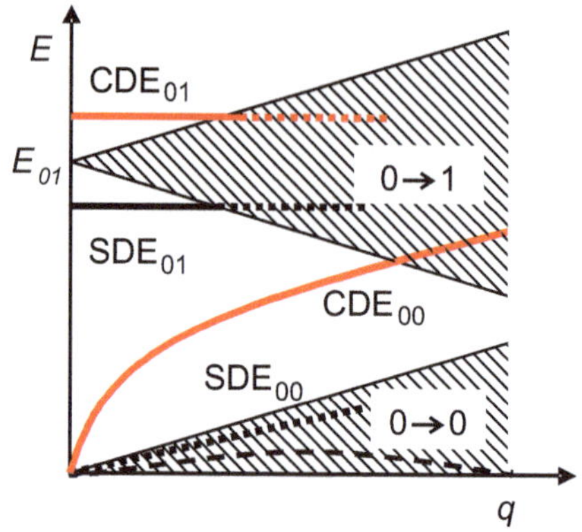

Fig. 3.6. Schematic picture of the dispersion of collective excitations of a Q2D system and the respective single–particle continua

Its diagram is topologically different from the diagram in (a). The transversal Coulomb interaction (Fig. 3.5(c)), which stems from the exchange of a virtual photon between the particles, is negligible, since it causes only very small energy renormalizations [5].

Now, we will try to match the two pictures, discussed so far: The independent–particle and the many–particle picture. Figure 3.6 displays schematically the single–particle continua, together with the dispersions of the collective CDE's and SDE's. The energies of CDE's are indicated by thick red (gray) lines, and those of SDE's by thick black lines. For clarity, we have labeled here the intersubband excitations by the subscript "01", and the intraband excitations by "00". Macroscopically, the intrasubband CDE_{00} and SDE_{00} can, like the intersubband excitations, be viewed as electron–density oscillations – but now *parallel* to the plane of the quantum well. Like the intrasubband SPE_{00}, the energy of the intrasubband CDE_{00} goes to zero as $q \rightarrow 0$. Collective modes – like the CDE's and SDE's – are defined eigenmodes of the electronic system for regions outside the single–particle continua, only. Inside the single–particle continua, the collective excitations are no longer eigenmodes of the system. There, they decay into uncorrelated electron–hole pairs, i.e., SPE's [1, 6, 7, 8, 9, 10]. This effect is called *Landau damping* [11]. It was invented in 1946 by L. D. Landau for the modes of liquid superfluid Helium. To notify this effect, in Fig. 3.6 the dispersions of the collective excitations are dotted inside the continua. From this we see in Fig. 3.6 that in particular the intrasubband SDE_{00} is Landau damped over the whole range of wave vectors q. We will discuss in detail the inelastic light scattering by these intra– and intersubband excitations of Q2D systems at the end of Chap. 4.

This section about the phenomenology of the elementary electronic excitations of a Q2D electron system shall be concluded by summarizing the explicit expressions for the energies of intersubband and intrasubband excitations. In a TDLDA theory, the energies of intersubband CDE_{01} and SDE_{01}

are given by [12, 13, 14, 15]

$$E^2_{CDE_{01}} = E^2_{01}\,(1 + \alpha_{11} + \beta_{11}) \tag{3.5}$$

$$E^2_{SDE_{01}} = E^2_{01}\,(1 + \beta_{11}) \tag{3.6}$$

α_{11} and β_{11} are terms, which stem from the direct and exchange part of the Coulomb interaction, respectively. In the following section we will derive explicit expressions for these contributions. The dispersion of intrasubband $\mathrm{CDE_{00}}$ was introduced for the first time by F. Stern [16] in 1967 for electrons on liquid Helium. It is given by

$$E^2_{CDE_{00}} = \frac{N_s \hbar^2 e^2}{2\,\epsilon_\infty \epsilon_0 m^*}\, q_\| \ , \tag{3.7}$$

where N_s is the carrier density of the 2DES, and ϵ_∞ the high-frequency dielectric constant[1]. Due to Landau damping, the energy of the intrasubband $\mathrm{SDE_{00}}$ is essentially given by the upper bound of the intrasubband single–particle continuum (see 3.1)

$$E^2_{SDE_{00}} = \frac{2\pi N_s}{m^{*2}}\, q_\|^2 \ . \tag{3.8}$$

3.3 Collective Excitations: Theoretical Models

In Sect. 2.4, we have discussed models to describe the ground state of Q2D electron systems. There, we have introduced the Hartree–, Hartree–Fock–, and the Kohn–Sham equations, which represent different levels of approximation. In order to determine the *dynamical* excitations of the system, one has to consider a time–dependent perturbation and then calculate the excitation spectrum self–consistently. The most commonly used approach to do so is the *random–phase approximation* (RPA). The RPA is a time–dependent Hartree approximation, i.e., the Coulomb interaction is treated classically and exchange–correlation effects are neglected. The energies of the collective excitations are determined by the poles of the frequency–dependent dielectric function, $\epsilon(\omega, \boldsymbol{q})$. Within the RPA, spin–density excitations are just single–particle excitations, since collective effects due to exchange interaction are neglected in this approximation. There are various improvements of the RPA, where, e.g., exchange–correlation effects are taken into account by the inclusion of a local LDA potential (see Sect. 2.4). At this level of approximation, the method is often called *time–dependent local–density approximation* (TDLDA), which is a time–dependent Kohn–Sham theory. If also the conservation of spin–current density is considered, a *time–dependent local–spin–density approximation* (TDLSDA) is formulated. The energies of

[1] For GaAs, $\epsilon_\infty = 10.9$.

the collective excitations are then, more generally, determined by the poles of a density–density correlation function, where, for CDE's the density in the correlation function is the electron density, and, for SDE's the spin density. If, on the other hand, the ground state is determined by a Hartree–Fock calculation, the excitations can be calculated employing a time–dependent Hartree–Fock theory. In later chapters we will derive various examples and comparisons of these theories with experiments by concrete low–dimensional semiconductor systems. In the literature, the majority of theories of electronic excitations in Q2D electron systems is performed on the basis of RPA or TDLDA[2].

3.3.1 Basic Ideas of RPA and TDLDA

In order to get a feeling how the time–dependent approximations work, we will briefly outline here the basic ideas of RPA, and then apply it to our two–subband system, modeling a Q2D electron system.

Let us consider an electromagnetic dipolar field

$$\boldsymbol{E}(t) = \boldsymbol{e}_z E_{ext} e^{-i\omega t} \tag{3.9}$$

in z direction. The single–particle Hamiltonian of the perturbed system can be written as

$$H = H_0 + H_1(t) \,, \tag{3.10}$$

with H_0 being the Hamiltonian of the electronic ground state, as, e.g., described in Sect. 2.4. The equation of motion for the corresponding statistical operator

$$\rho = \rho_0 + \rho_1(t) \tag{3.11}$$

is the von Neumann equation

$$\frac{d}{dt}\rho = -\frac{i}{\hbar}\left[H, \rho\right] \,. \tag{3.12}$$

Inserting (3.11) into (3.12), and neglecting the term $\rho_1 H_1$ yields

$$i\hbar\frac{d}{dt}\rho = [H_1(t), \rho_0] + [H_0, \rho_1(t)] \,. \tag{3.13}$$

We now make the assumption that $H_1(t)$ and $\rho_1(t)$ have the same temporal dependence as the external perturbation, i.e.,

$$H_1(t) = H_1 e^{-i\omega t} \,, \tag{3.14}$$

and,

[2] For RPA calculations, see, e.g., [16, 17, 18, 19, 20, 21, 22, 23, 24, 25, 26], for TDLDA [12, 13, 14, 15], for TDLSDA [27], and, for TD–HFA [28].

$$\rho_1(t) = \rho_1 e^{-i\omega t} . \tag{3.15}$$

Let $\{|n\rangle\}$ be the known eigenstates of the ground–state Hamiltonian H_0. Then, we get with (3.13) through (3.15)

$$\begin{aligned} \hbar\omega\langle m|\rho_1|n\rangle &= \langle m|\left[H_1,\rho_0\right]|n\rangle + \langle m|\left[H_0,\rho_1\right]|n\rangle = \\ &= (f_n - f_m)\,\langle m|H_1|n\rangle + (E_m - E_n)\,\langle m|\rho_1|n\rangle , \end{aligned} \tag{3.16}$$

with f_n being the Fermi distribution function at the energetic position E_n. From (3.16) it follows that

$$\langle m|\rho_1|n\rangle = \frac{f_n - f_m}{E_n - E_m + \hbar\omega}\langle m|H_1|n\rangle . \tag{3.17}$$

The perturbation causes an induced density, $\delta n(\boldsymbol{r})$, which is given by

$$\begin{aligned} \delta n(\boldsymbol{r}) &= Tr\left\{\delta(\boldsymbol{r}-\boldsymbol{r}')\rho_1\right\} = \sum_n \langle n|\delta(\boldsymbol{r}-\boldsymbol{r}')\rho_1|n\rangle = \\ &= \sum_{m,n}\langle n|\delta(\boldsymbol{r}-\boldsymbol{r}')|m\rangle\langle m|\rho_1|n\rangle = \\ &= \sum_{m,n}\langle m|\rho_1|n\rangle\psi_n^*(\boldsymbol{r})\psi_m(\boldsymbol{r}) . \end{aligned} \tag{3.18}$$

Inserting (3.17) into (3.18) delivers the general result for the induced density

$$\delta n(\boldsymbol{r}) = \sum_{m,n}\frac{f_n - f_m}{E_n - E_m + \hbar(\omega + i\eta)}\psi_n^*(\boldsymbol{r})\psi_m(\boldsymbol{r})\langle m|H_1|n\rangle , \tag{3.19}$$

where we have introduced a phenomenological damping term $i\eta$.

3.3.2 Application to Two–Subband System

Now, we will apply this result to our Q2D two–subband system, i.e., in the following we have to consider the z direction, only. For the perturbation operator H_1 we have

$$H_1 = V_{ext} + \Delta V_{Coul} + \Delta V_{XC} , \tag{3.20}$$

with the classical or direct term of the Coulomb interaction (Hartree term)

$$\Delta V_{Coul} = -\frac{e^2}{\epsilon_\infty\epsilon_0}\int_{-\infty}^{z} dz' \int_{-\infty}^{z'} dz''\delta n(z'') , \tag{3.21}$$

and the exchange–correlation term

$$\Delta V_{XC}(z) = \frac{\partial V_{XC}}{\partial n}\delta n(z) . \tag{3.22}$$

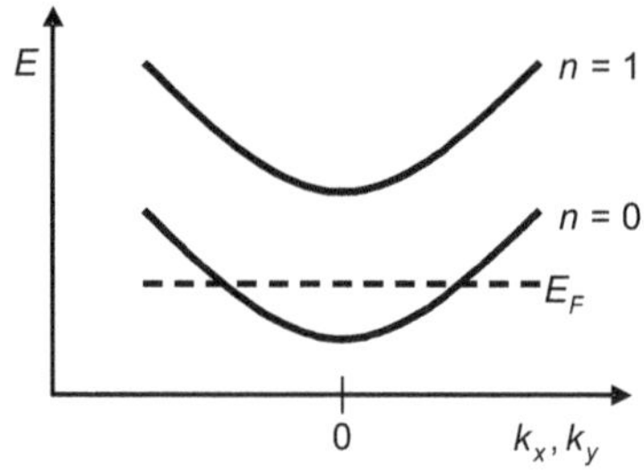

Fig. 3.7. Schematic picture of a Q2D electron system with two subbands. The label k_x, k_y at the x axis shall indicate that for a Q2D system the dispersion of a subband is not a one–dimensional parabola but rather a paraboloid in k_x and k_y directions

V_{XC} is the statical exchange–correlation potential, taken from LDA [29]. If we now consider vertical ($q = 0$) intersubband excitations of the two–subband system, as displayed in Fig. 3.7, only, we get from (3.19)

$$\begin{aligned}\delta n(z) &= N_s\left(\frac{1}{E_0 - E_1 + \hbar\omega} - \frac{1}{E_1 - E_0 + \hbar\omega}\right)\chi_0(z)\chi_1(z)\langle 1|H_1|0\rangle = \\ &= 2N_s\frac{E_{01}}{(\hbar\omega)^2 - E_{01}^2}\chi_0(z)\chi_1(z)\langle 1|H_1|0\rangle \,. \end{aligned} \tag{3.23}$$

Here, the $\chi_i(z)$ are the envelope functions of the quantum well in z direction (cf. Subsect. 2.4.1). Finally, inserting (3.23) into (3.20) yields the self–consistency equations for the dynamical potential

$$\langle 1|H_1|0\rangle = \langle 1|V_{ext}|0\rangle + \frac{E_{01}^2}{(\hbar\omega)^2 + E_{01}^2}\left(\alpha_{11} + \beta_{11}\right)\langle 1|H_1|0\rangle \,. \tag{3.24}$$

In (3.24), the direct and exchange–correlation terms, α_{11} and β_{11}, are given by

$$\alpha_{11} = -\frac{2e^2 N_s}{\epsilon_\infty\epsilon_0 E_{01}}\int_{-\infty}^{z} dz' \int_{-\infty}^{z'} dz''\chi_1(z'')\chi_0(z'') \,, \tag{3.25}$$

$$\beta_{11} = \frac{2N_s}{E_{01}}\int_{-\infty}^{\infty} dz \frac{\partial V_{XC}}{\partial n}\chi_1^2(z)\chi_0^2(z) \,, \tag{3.26}$$

respectively. Characteristic for the collective modes is that even for a vanishing external perturbation ($V_{ext} = 0$), a nontrivial solution of (3.24) exists. This is equivalent to the condition

$$\epsilon(\omega, \boldsymbol{q}) = 0 \,. \tag{3.27}$$

Hence, we arrive with (3.24) at

$$E_{CDE_{01}}^2 \equiv (\hbar\omega)^2 = E_{01}^2\left(1 + \alpha_{11} + \beta_{11}\right) \tag{3.28}$$

for the energy of the intersubband CDE_{01}. Without proof, the energy of the corresponding SDE_{01} is given by

$$E^2_{SDE_{01}} = E^2_{01}\,(1+\beta_{11}) \ . \tag{3.29}$$

With a treatment, as done here, we can not justify SDE's, since they are not dipole active, i.e., they can not be created by a dipolar electromagnetic field. We have already discussed this in the previous subsection on a phenomenological basis. The excitation of SDE's in resonant inelastic light scattering experiments is possible due to spin–orbit coupling of the valence–band states, which serve as intermediate states in the scattering process. We will come back to this important – but more involved – point in the next section, when we discuss the scattering mechanisms. For the quantities α_{11} and β_{11}, which describe the effects of direct and exchange–correlation interactions, respectively, the relations

$$\alpha_{11} > 0\,, \quad \beta_{11} < 0 \tag{3.30}$$

hold. For electron densities, N_s, in the range 10^{11} cm^{-2}, also

$$|\alpha_{11}| > |\beta_{11}| \ . \tag{3.31}$$

Therefore, in most practical cases, the CDE_{01} is blueshifted, and, the SDE_{01} is redshifted with respect to the corresponding SPE_{01} (cf. Fig. 3.3). As an illustration, Fig. 3.8 displays the energies of the collective intersubband excitations and of the bare subband spacing E_{01} for a one–sided doped, 25 nm–wide GaAs-AlGaAs quantum well in dependence on the carrier density

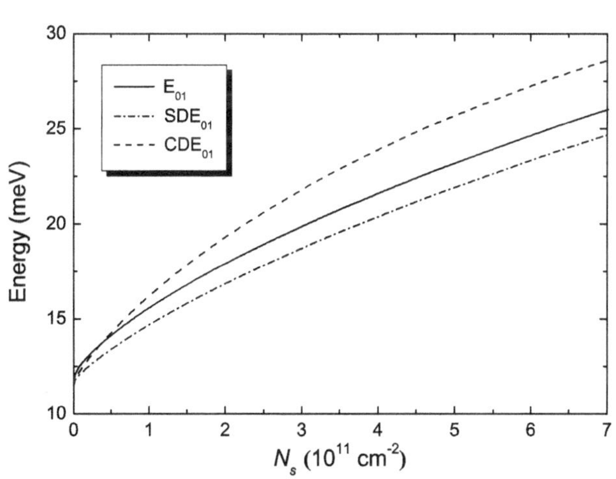

Fig. 3.8. Calculated Energies of collective intersubband excitations, compared to the bare subband spacing, E_{01}, in a one–sided doped GaAs–AlGaAs quantum well (data after [30])

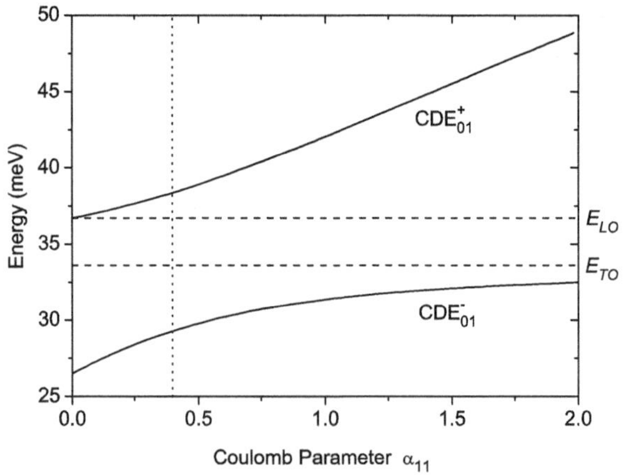

Fig. 3.9. Coupled intersubband plasmon–LO phonon modes, calculated with (3.33) for a bare subband spacing of $E_{01} = 26.5$ meV

N_s, calculated within TDLDA. One can see that at very small N_s, exchange–correlation effects become dominant, and $|\alpha_{11}| < |\beta_{11}|$ (see also [31], and references therein).

3.3.3 Plasmon–LO Phonon Coupling

Another interesting point is that in polar semiconductors, like GaAs, the collective CDE's can couple – via their macroscopic electric field – to the polar LO phonons. To take this coupling into account, we have to replace in the term of the direct Coulomb interaction, (3.25), the high–frequency dielectric constant, ϵ_∞, by the frequency–dependent dielectric function of the host lattice, $\epsilon_L(\omega)$. If we use the dielectric function for polar bulk semiconductors,

$$\epsilon_L(\omega) = \epsilon_\infty \frac{E_{LO}^2 - (\hbar\omega)^2}{E_{TO}^2 - (\hbar\omega)^2} , \tag{3.32}$$

with E_{LO} and E_{TO} being the longitudinal and transverse optical phonon energies[3], respectively, we get two coupled LO phonon–plasmon modes from (3.28) (see, e.g., [20, 22, 32, 33, 34]):

$$E_\pm = \left\{ \frac{1}{2} \left[E_{LO}^2 + E_{01}^2 (1 + \alpha_{11}) \right] \pm \right.$$
$$\left. \pm \frac{1}{2} \sqrt{\left[E_{LO}^2 + E_{01}^2 (1 + \alpha_{11}) \right]^2 - 4E_{01}^2 \left[E_{LO}^2 + E_{TO}^2 \alpha_{11} \right]} \right\}^{1/2} . \tag{3.33}$$

For illustration, Fig. 3.9 shows the energies of the two coupled intersubband plasmon–LO phonon modes, calculated with (3.33), for a subband spacing of

[3] For GaAs, $E_{LO} = 36.7$ meV, and, $E_{TO} = 33.6$ meV.

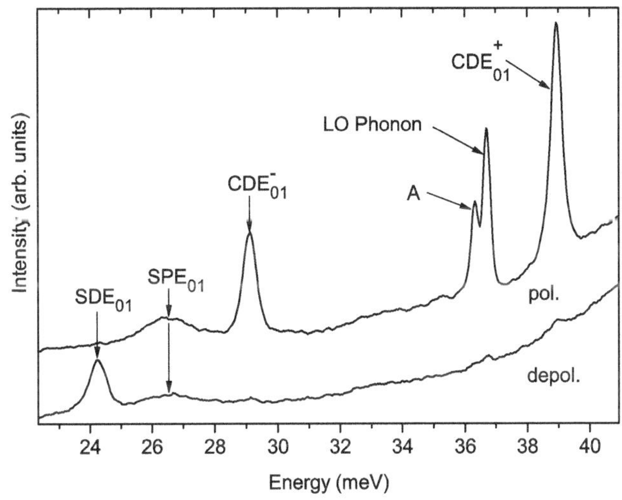

Fig. 3.10. Polarized and depolarized spectra of intersubband excitations in a 25 nm GaAs–AlGaAs quantum well with carrier density $N_s = 6.5 \times 10^{11}$ cm^{-2}

$E_{01} = 26.5$ meV in dependence on the parameter α_{11} of the direct Coulomb interaction. Indeed, for the experimental spectra in Fig. 3.3, the displayed energy range is chosen just around the single–particle spacing, E_{01}, and the observed intersubband CDE is actually the lower one of the two coupled modes. Figure 3.10 shows spectra of intersubband excitations of a similar quantum well over a larger range of energies. One can clearly observe the coupled intersubband plasmon–LO phonon modes CDE$^-_{01}$ and CDE$^+_{01}$. The sharp line at the energy of the LO phonon is due to a signal from pure phonon scattering in the GaAs buffer layer of the sample. The line labeled "A" is a coupled mode corresponding to an intersubband excitation to the next higher (the third) subband. The vertical dotted line in Fig. 3.9 indicates the positions of the modes, observed experimentally in Fig. 3.10. There is a reasonable agreement with experiment, within this simplified two–subband model, where coupling to higher transitions is neglected.

References

1. A. Pinczuk, S. Schmitt–Rink, G. Danan, J. P. Valladares, L. N. Pfeiffer, and K. W. West: Phys. Rev. Lett. **63**, 1633 (1989)
2. A. Pinczuk, H. L. Störmer, R. Dingle, J. M. Worlock, W. Wiegmann, and A. C. Gossard: Solid State Commun. **32**, 1001 (1979)
3. G. Abstreiter and K. Ploog: Phys. Rev. Lett. **42**, 1308 (1979)
4. E. Batke, G. Weimann, W. Schlapp: Phys. Rev. B **43**, 6812 (1991)
5. M. Y. Jiang: Solid State Commun. **84**, 81 (1992)
6. L. Wendler: Solid State Commun. **65**, 1197 (1988)

7. D. Gammon, B. V. Shanabrook, J. C. Ryan, D. S. Katzer: Phys. Rev. B **41**, 12311 (1990)
8. D. Gammon, B. V. Shanabrook, J. C. Ryan, G. Brozak, D. S. Katzer: Surf. Sci. **263**, 471 (1992)
9. D. Gammon, B. V. Shanabrook, J. C. Ryan, D. S. Katzer, M. J. Yang: Phys. Rev. Lett. **68**, 1884 (1992)
10. Hong Yu, J. C. Hermanson: Phys. Rev. B **43**, 4340 (1991)
11. L. D. Landau: J. Phys. (USSR) **10**, 25 (1946)
12. T. Ando, A. B. Fowler, F. Stern: Rev. Mod. Phys. **54**, 437 (1982)
13. T. Ando, S. Katayama: J. Phys. Soc. Jpn. **54**, 1615 (1985)
14. G. Eliasson, P. Hawrylak, J. J. Quinn: Phys. Rev. B **35**, 5569 (1987)
15. A. C. Tselis, J. J. Quinn: Phys. Rev. B **29**, 3318 (1984)
16. F. Stern: Phys. Rev. Lett. **18**, 546 (1967)
17. J. K. Jain, S. Das Sarma: Phys. Rev. B **36**, 5949 (1987)
18. L. Wendler, R. Pechstedt: Phys. Rev. B **35**, 5887 (1987)
19. L. Wendler, R. Pechstedt: phys. stat. sol. (b) **141**, 129 (1987)
20. G. Abstreiter, R. Merlin, A. Pinczuk: IEEE J. Quantum Electron. **22**, 1771 (1986)
21. K. Bajema, R. Merlin, F. Y. Young, S. C. Hong, J. Singh, P. K. Bhattacharya: Phys. Rev. B **36**, 1300 (1987)
22. E. Burstein, A. Pinczuk, D. L. Mills: Surf. Sci. **98**, 451 (1980)
23. D. Y. Oberli, D. R. Wake, M. V. Klein, J. Klem, T. Henderson, H. Morkoc: Phys. Rev. Lett. **59**, 696 (1987)
24. A. Pinczuk, J. M. Worlock: Surf. Sci. **113**, 69 (1982)
25. D. A. Dahl, L. J. Sham: Phys. Rev. B **16**, 651 (1977)
26. S. J. Allen Jr., D. C. Tsui, B. Vinter: Solid State Commun. **20**, 425 (1976)
27. O. Steffens and M. Suhrke: Phys. Rev. Lett. **82**, 3891 (1999)
28. C. Steinebach, C. Schüller, G. Biese, and D. Heitmann: Phys. Rev. B **57**, 1703 (1998)
29. O. Gunnarson and B. I. Lundqvist: Phys. Rev. B **13**, 4274 (1976)
30. Christoph Steinebach: *Selbstkonsistente Berechnung des Potentialverlaufs und der optischen Anregungen in GaAs/AlGaAs-Heterostrukturen* Diplomarbeit, Universität Hamburg 1996, p. 13
31. S. Ernst, A. R. Goñi, K. Syassen, and K. Eberl: Phys. Rev. Lett. **72**, 4029 (1994)
32. S. DasSarma: Surface Sci. **11/12**, 535 (1982)
33. A. Pinczuk, J. M. Worlock, H. L. Störmer, R. Dingle, W. Wiegmann, and A. C. Gossard: Solid State Commun. **36**, 43 (1980)
34. A. Pinczuk and J. M. Worlock: Surface Sci. **113**, 69 (1982)

4 Basic Concepts of Inelastic Light Scattering, Experiments on Quantum Wells

4.1 Macroscopic Approach

4.1.1 General Remarks

Inelastic light scattering – or Raman scattering – is the scattering of light by a medium, where in the scattering process excitations are created (Stokes process) or annihilated (Antistokes process) within the medium. In solids, these excitations can be various types of elementary excitations, like phonons, magnons, or – as considered in this book – electronic excitations. Historically, the scattering by optical phonons, or by internal vibrations of molecules, is called Raman scattering, and the scattering by acoustic phonons Brillouin scattering. For *electronic* Raman scattering, often the term *inelastic light scattering* is used, so in this book. Figure 4.1 shows schematically the Stokes and Antistokes processes, where a photon with energy $\hbar\omega_I$ and momentum $\boldsymbol{k}_I$ is scattered by the creation or annihilation of an elementary excitation with energy $\hbar\omega$ and momentum $\boldsymbol{q}$. The scattered photon has an energy $\hbar\omega_S$ and momentum $\boldsymbol{k}_s$. This means, each scattered photon in the Stokes component is associated with a gain in energy $\hbar\omega$ by the sample. Similarly, the sample loses energy $\hbar\omega$ for each scattered photon in the Antistokes component

$$\hbar\omega_S = \hbar\omega_I \pm \hbar\omega \; , \tag{4.1}$$

where the minus (plus) sign is for the Stokes (Antistokes) process. Conservation of momentum requires

$$\boldsymbol{k}_S = \boldsymbol{k}_I \pm \boldsymbol{q} \; . \tag{4.2}$$

Here, the plus (minus) sign is for the Stokes (Antistokes) process. Schematically, a Raman spectrum of a single excitation with energy $\hbar\omega$ looks like the one displayed in Fig. 4.2. There is a very strong component from elastic scattering at energy $\hbar\omega_I$, e.g., from the surface of the sample. For experiments at low temperatures, as considered in this book for experiments on semiconductor nanostructures, only the Stokes process has a significant probability. Therefore, all experiments presented in this book are exclusively measurements of the Stokes components. Usually, the energy axis of the experimental

Christian Schüller: *Inelastic Light Scattering of Semiconductor Nanostructures*
STMP **219**, 57–84 (2006)
DOI 10.1007/3-540-36526-5_4

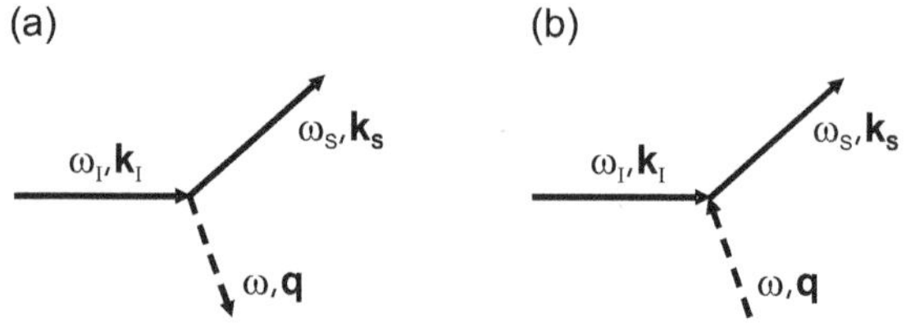

Fig. 4.1. Schematic picture of (**a**) the Stokes, and, (**b**) the Antistokes scattering process

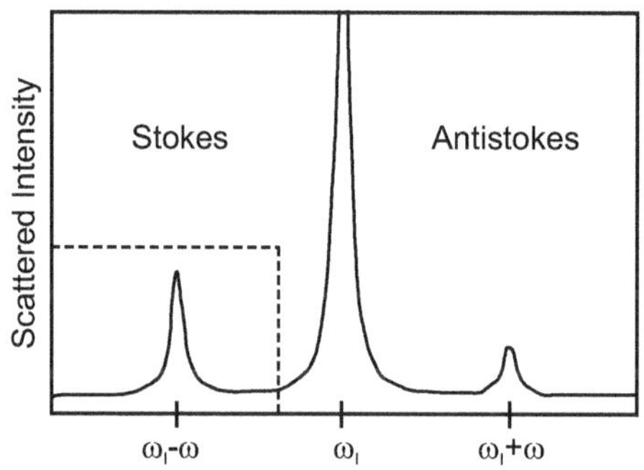

Fig. 4.2. Schematic picture of a Raman spectrum

spectra are shifted so that the origin is at the position of the energy of the incident photons, $\hbar\omega_I$.

In semiconductors, the energies of elementary excitations are always small compared to the energy of the incident laser photons, i.e., $\hbar\omega \ll \hbar\omega_I$. A particular strength of the method therefore is that elementary excitations with energies in the far–infrared spectral range can be measured in the visible range ($\hbar\omega_I$, $\hbar\omega_S$), where powerful lasers and detectors are available. A further strength of the method is the possibility to transfer a finite quasimomentum, or wave vector, $\boldsymbol{q}$, to the excitation during the scattering process. The maximum wave vector can be transferred employing the exact backscattering geometry, i.e., the directions of incoming and scattered light are antiparallel. For this geometry, the maximum value, $q_{\max}$, of the transferred momentum is twice the momentum of the light, under the assumption that the wavelengths of the incoming, λ_I, and the scattered photons, λ_S, are approximately equal:

$$q_{\max} = \frac{4\pi}{\lambda_I} \, . \tag{4.3}$$

For quantum wells, translational invariance is only valid within the plane of the well. In the perpendicular direction, i.e., the growth direction, the translational symmetry is not conserved. Therefore, in experiments on quantum wells, a quasi–continuously tunable momentum transfer is possible in the lateral directions, only. Figure 4.3 shows a schematic picture of the

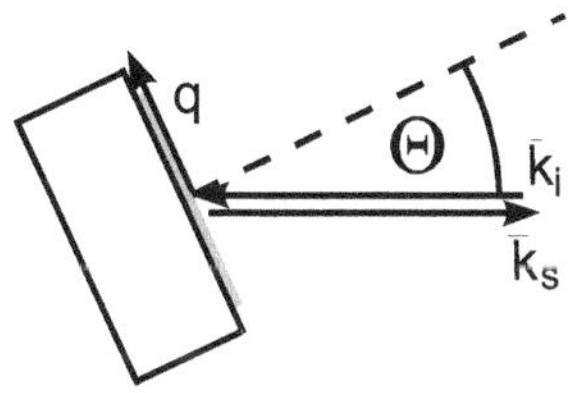

Fig. 4.3. Schematic picture of the backscattering geometry for experiments on quantum wells

backscattering geometry, which is typically used for experiments on semiconductor quantum wells. The wave–vector transfer q, parallel to the plane of the well, is for this case given by

$$q = \frac{4\pi}{\lambda_I} \sin\Theta \,, \tag{4.4}$$

provided that $\lambda_I \approx \lambda_S$. If generally the angles Θ_I and Θ_S of the incident and scattered beams with respect to the sample normal are not equal, the wave–vector transfer q is given by

$$q = \frac{2\pi}{\lambda_I} \left[\sin(\Theta_I) + \sin(\Theta_S)\right] \,. \tag{4.5}$$

For experiments on wires, a quasi–continuously tunable momentum can only be transferred parallel to the wire direction. Perpendicular to the wires, electron motion, and hence momentum, is quantized. For quantum dots, the latter holds for all lateral directions. We will come to the specifics of those structures in the respective chapters about experiments on wires and dots later in this book.

4.1.2 Macroscopic Point of View

On general grounds, in order for inelastic scattering to take place, the medium must have fluctuations in time and space in some system coordinates, which we in the following will call $Q(\boldsymbol{r}, t)$:

$$Q(\boldsymbol{r}, t) = \sum_{\boldsymbol{q}} \left\{Q(\boldsymbol{q}, t)\exp(i\boldsymbol{q}\boldsymbol{r}) + Q^*(\boldsymbol{q}, t)\exp(-i\boldsymbol{q}\boldsymbol{r})\right\} \,. \tag{4.6}$$

In a completely homogeneous medium, there is no inelastic light scattering. These fluctuations can, e.g., be lattice displacements, or electron–density fluctuations. The former would lead to scattering by phonons and the latter to scattering by electronic excitations.

Macroscopically, the inelastic scattering of light can be described as the scattering by an extended medium, where the individual dipole moments combine to form a macroscopic polarization vector $\boldsymbol{P}(\boldsymbol{r},t)$ (see, e.g., [1], and references therein) with

$$\boldsymbol{P}(\boldsymbol{r},t) = \epsilon_0 \chi_0 \boldsymbol{E}_I(\boldsymbol{r},t) , \tag{4.7}$$

where the susceptibility, χ_0, is generally a tensor. $\boldsymbol{E}_I(\boldsymbol{r},t)$ is the electric field of the incident light. The scattered beam is radiated by the oscillatory macroscopic polarization, which is subject to the usual Maxwell equations. Quantum mechanically, the elementary excitations of the medium modulate the wave functions and energy levels of the medium. The changes in these quantities are linear in $Q(\boldsymbol{r},t)$ in first order perturbation theory. Their effect is represented macroscopically by an additional contribution to the susceptibility, which we write symbolically (since χ is generally a tensor)

$$\boldsymbol{P}(\boldsymbol{r},t) = \epsilon_0 \chi(\omega_I) \boldsymbol{E}_I(\boldsymbol{r},t) = \epsilon_0 \left(\chi_0 + \frac{\partial \chi}{\partial Q} Q + \dots \right) \boldsymbol{E}_I(\boldsymbol{r},t) . \tag{4.8}$$

χ_0 oscillates at the same frequency as the incident field and contributes to elastic scattering, only. The second expression in (4.8), $\frac{\partial \chi}{\partial Q} Q$, is a second–order susceptibility, which describes the modulation. The second–order polarization resulting from this term oscillates at frequencies different from the frequency ω_I of the incident light. This is because Q itself is a time–dependent function. One can show that the second–order polarization can be separated into two parts: the Stokes– and the Antistokes part (see, e.g., [1]).

The key quantity of Raman scattering is the spectral differential cross section $\frac{d^2\sigma}{d\Omega d\omega_S}$. For the Stokes part of the spectrum it is, for a fixed scattering angle, determined as the rate of removal of energy from the incident beam as a result of its scattering into a solid–angle element $d\Omega$ with frequency between ω_S and $\omega_S + d\omega_S$, divided by the product of $d\Omega d\omega_S$ with the incident beam intensity. In the macroscopic theory, the spectral differential cross section is given by

$$\frac{d^2\sigma}{d\Omega d\omega_S} \propto \left| \boldsymbol{e}_S \frac{\partial \chi}{\partial Q} \boldsymbol{e}_I \right|^2 \langle Q^*(\boldsymbol{q}) Q(\boldsymbol{q}) \rangle_\omega . \tag{4.9}$$

The quantity $\frac{\partial \chi}{\partial Q}$ is the so called Raman tensor. The first factor in (4.9) describes, e.g., polarization selection rules due to the product of the Raman tensor with the polarizations $\boldsymbol{e}_I$ and $\boldsymbol{e}_S$ of the incident and scattered light. We will elaborate on this product in detail in the next sections, when we consider the microscopic theory, to deduce the polarization selection rules for inelastic light scattering. This term is also responsible for the strength of the scattering, i.e., the signal intensity, since it describes resonance effects: For resonant scattering, i.e., if the energy of the incident photons is close to a real transition (e.g., valence band to conduction band transition) in the crystal, this term can become very large. The second term in (4.9) is the power spectrum of the fluctuations $Q(\boldsymbol{r},t)$. It determines the shape of the spectrum, i.e., where peaks occur due to scattering processes. The power spectrum is defined via the relation

$$\langle Q^*(\boldsymbol{q},\omega)Q(\boldsymbol{q},\omega')\rangle = \langle Q^*(\boldsymbol{q})Q(\boldsymbol{q})\rangle_\omega \delta(\omega-\omega') \,. \tag{4.10}$$

The brackets denote an average over the probability distribution. The two quantities in the brackets on the left–hand side of (4.10) are independent random variables whose phases can take all values between 0 and 2π. Their product therefore has a zero average, except for $\omega = \omega'$. The power spectrum plays an important role in the macroscopic theory of inelastic light scattering. Equation (4.9) shows that the cross section for scattering by an excitation is proportional to its power spectrum.

4.1.3 Dissipation–Fluctuation Analysis

An important further analysis in the theory of inelastic light scattering – in the macroscopic as well as in the microscopic versions (see also below) – is the application of the dissipation–fluctuation theorem:

$$\langle Q^*(\boldsymbol{q},t)Q(\boldsymbol{q},0)\rangle_\omega = \frac{1}{2\pi}\int_{-\infty}^{\infty} e^{i\omega t}\langle Q^*(\boldsymbol{q},t)Q(\boldsymbol{q},0)\rangle dt \tag{4.11}$$

$$= \frac{\hbar}{\pi}\left(n_\omega+1\right)\Im\left\{\Pi(\boldsymbol{q},\omega)\right\}, \tag{4.12}$$

where the quantities $Q(\boldsymbol{q},t)$ are the above introduced Fourier transforms, $Q(\boldsymbol{q})$, of the fluctuations in the time–dependent Heisenberg picture. Equation (4.12) is the application of the dissipation–fluctuation theorem: It states that the power spectrum of the fluctuations is proportional to the imaginary part of the response function, $\Pi(\boldsymbol{q},\omega)$, of the system to the fluctuations. n_ω is the well–known Bose–Einstein factor for bosons.

With (4.9) and (4.12) we see that the cross section for inelastic light scattering is proportional to the imaginary part of the response function of the system to the fluctuations, which cause the inelastic scattering:

$$\frac{d^2\sigma}{d\Omega d\omega_S} \propto \left(n_\omega+1\right)\Im\left\{\Pi(\boldsymbol{q},\omega)\right\} \,. \tag{4.13}$$

In the case of scattering by electronic excitations, the response function $\Pi(\boldsymbol{q},\omega)$ is the density–density correlation function, which will be discussed in detail in the following sections. It can also be shown that

$$\Im\left\{\Pi(\boldsymbol{q},\omega)\right\} \propto \Im\frac{-1}{\epsilon\left(\boldsymbol{q},\omega\right)}, \tag{4.14}$$

with $\epsilon\left(\boldsymbol{q},\omega\right)$ being the wave–vector and frequency–dependent dielectric function of the system. In most of the theoretical work, which can be found in literature on electronic excitations in semiconductor nanostructures, the expression (4.13) is employed and the density–density correlation function is calculated assuming nonresonant conditions, i.e., transitions between valence– and conduction–band states are neglected. In this situation, which is of course much less complex than including the valence band, the density–density correlation function for the electron system in the conduction band is calculated, only. In the remainder of this book we will see that for realistic experimental conditions this approximation is not appropriate, since valence–band states play an important role in resonant inelastic light scattering experiments.

4.2 Microscopic Approach, Polarization Selection Rules

In this section, the dominant *microscopic* scattering mechanisms, which lead to the light scattering in low–dimensional electron systems, are discussed. In our work we found that in particular a second–order *resonant* scattering process plays a crucial role for the observation of single–particle–like excitations in quantum wells, wires, and dots [2, 3]. For illustration, we will qualitatively discuss in the first subsection the different scattering mechanisms by the example of Q2D intersubband excitations in quantum wells, which were introduced in Sect. 3.8. In the following two subsections, a more general and profound discussion of the scattering cross section for scattering by electron–density fluctuations will be given and polarization selection rules will be derived.

4.2.1 Two- and Three-Step Scattering Processes

In general, the inelastic light scattering by electronic excitations in semiconductor microstructures with relatively small numbers of electrons can be observed under specific interband resonance conditions, only. This means that in the scattering process valence–band states are involved as intermediate states. Nevertheless, most of the theoretical work concerning the dynamic Raman response in microstructures is performed for nonresonant conditions, neglecting the valence–band structure. We summarize here the dominant *resonant* scattering processes, which are fragmentarily present in the literature, and which we found in our experiments on quantum wells, wires, and dots to be the dominant ones [3].

Quantum mechanically, the general microscopic problem is to describe the transition of a photon from the state $(\omega_I, \boldsymbol{k}_I, \boldsymbol{e}_I)$ – with frequency ω_I, quasimomentum $\boldsymbol{k}_I$, and polarization $\boldsymbol{e}_I$ – to the state $(\omega_S, \boldsymbol{k}_S, \boldsymbol{e}_S)$ with changed quantities, while the many–particle system undergoes a transition from the many–particle initial state $|I\rangle$ to the final state $|S\rangle$. For such a scattering event, the spectral differential cross section is generally given by [4]

$$\frac{d^2\sigma}{d\Omega d\omega_S} = \hbar \frac{\omega_S}{\omega_I} < \sum_S |M_{SI}|^2 \, \delta(\hbar\omega_I - \hbar\omega_S - \hbar\omega) >, \tag{4.15}$$

where M_{SI} is the matrix element, which describes the transition of the system from state $|I\rangle$ to state $|S\rangle$ via the scattering process, and the sum runs over all possible final states. Of course, in most cases the exact many–particle states $|I\rangle$ and $|S\rangle$ are not known, and approximations have to be applied.

The coupling of the radiation with the electron system is taken into account by replacing the momentum $\boldsymbol{p}$ of the electron by $\boldsymbol{p}+e\boldsymbol{A}$ in the Hamiltonian H_0 of the unperturbed system. $\boldsymbol{A}$ is the sum of the vector potentials of the incident and scattered electromagnetic fields.

$$H = \frac{1}{2m}\sum_i \left[(\boldsymbol{p}_i + e\boldsymbol{A}(\boldsymbol{r}_i))^2 + U(\boldsymbol{r}_i)\right] + \hat{V}_{e-e} + \hat{V}_{e-ph} = \\ = H_0 + \frac{1}{2m}\sum_i \left[(\boldsymbol{A}(\boldsymbol{r}_i))^2 + \boldsymbol{p}_i\boldsymbol{A}(\boldsymbol{r}_i) + \boldsymbol{A}(\boldsymbol{r}_i)\boldsymbol{p}_i\right] \quad (4.16)$$

$\hat{V}_{e-e}$ is the Coulomb interaction, $\hat{V}_{e-ph}$ the electron–phonon interaction, and U includes the lattice–periodic potential as well as all types of external potentials. For simplicity, the spin–orbit coupling is neglected in (4.16). In a perturbation theory, the last three terms in the second line of (4.16) can be treated as a perturbation. Here the $\boldsymbol{A}^2$ terms give contributions to the light scattering cross section in *first* order, the $\boldsymbol{pA}$ terms in *second* order perturbation theory. The $\boldsymbol{pA}$ terms together with, either the Coulomb interaction $\hat{V}_{e-e}$ of a photoexcited exciton with the Fermi sea, or the electron–phonon interaction $\hat{V}_{e-ph}$ of a photoexcited exciton with the phonon bath, give contributions in *third* order perturbation theory. $\hat{V}_{e-ph}$ leads to the scattering by phonons which shall not be discussed here. The second–order (SOP) and third–order light scattering processes (TOP) appear to be the dominant ones in electronic Raman scattering on microstructures because they exhibit a resonant behavior. As an example for nanostructured systems, we sketch in Fig. 4.4 transitions which contribute to the second (Fig. 4.4 (a), (b) [6]) and

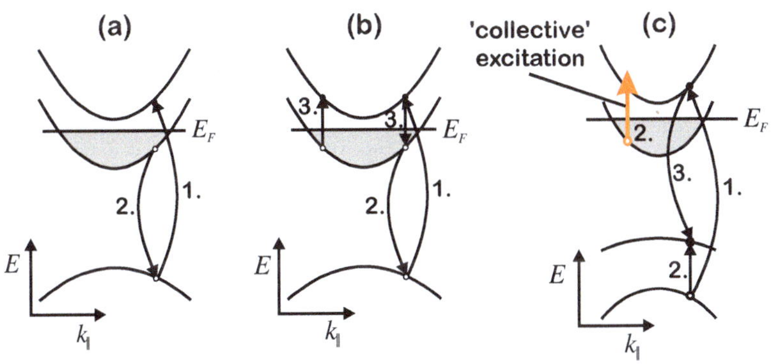

Fig. 4.4. Schematic picture of the (**a**) two–step and (**b**) three–step second–order scattering mechanisms, and (**c**) the three–step third–order excitonic scattering mechanism for inelastic light scattering by electronic excitations in quantum wells

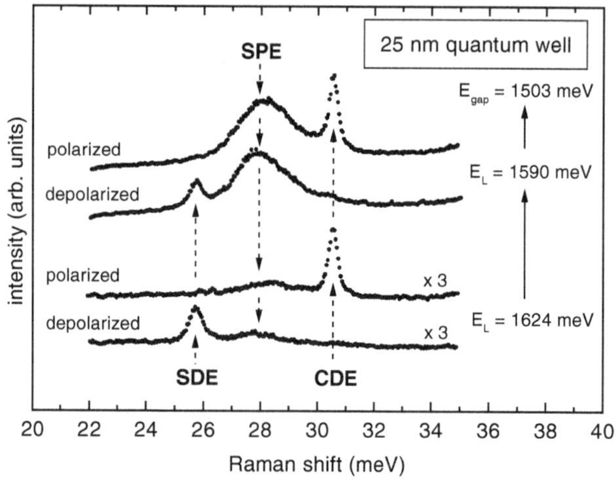

Fig. 4.5. Depolarized and polarized Raman spectra of Q2D intersubband excitations for two different laser energies E_L, and enhancement of the SPE for excitation under extreme–resonance conditions

third–order (Fig. 4.4 (c) [7]) processes in the scattering by intersubband excitations in quantum wells in backscattering geometry ($\boldsymbol{q} = 0$). We have found in our investigations that qualitatively these scattering processes also apply to Q1D and Q0D systems [2, 3].

In the first step of the two–step SOP, an electron is excited from a valence–band to a conduction–band state. In the second step, an electron, either with the same or with opposite spin as the photo–excited electron, recombines with the hole in the valence band. Thus, as a net effect, either a nonspinflip or spinflip single–particle excitation has been created in the conduction band. In the third step of the three–step SOP, another SPE is created by Coulomb interaction. This represents the screening of the SPE by other SPE's in the system. This screening also leads to the formation of collective CDE's and SDE's by direct and exchange Coulomb coupling. For nonresonant conditions, the SPE's are screened by the interaction. This is formally implemented by the cancellation of the two–step SOP contribution by the three–step SOP in the scattering cross section. We found in our experiments that under conditions of extreme resonance, which means that the laser frequency is in the vicinity of the fundamental bandgap of the underlaying Q2D structure, unscreened SPE's occur and that they show stronger resonance enhancements than the collective excitations [2, 3]. This is demonstrated in Fig. 4.5 by the Q2D intersubband excitations. The lower spectra are recorded at a laser energy E_L well above the effective bandgap ($E_{gap} = 1503$ meV). These spectra look very similar to the spectra reported by A. Pinczuk et al. in high–mobility quantum wells [8] (see also discussion of Fig. 3.3 in Sect. 3.8). What we found is that if we lower the laser energy towards the effective bandgap (upper spectra in

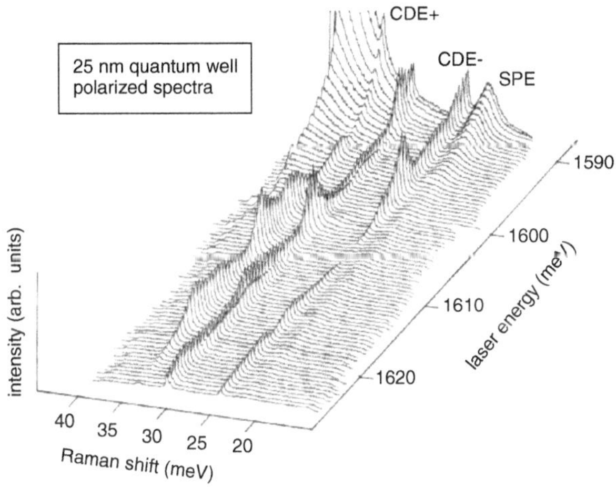

Fig. 4.6. Polarized Raman spectra of Q2D intersubband excitations for different laser energies E_L. The increasing background for $E_L < 1590$ meV is due to luminescence at the E_0 gap

Fig. 4.5), the SPE exhibits a stronger resonance enhancement than the collective excitations. This means that the above mentioned cancellation of terms does not hold in the case of extreme resonance and thus single–particle–like excitations[1] can be observed. A quite similar behavior was previously reported for intraband excitations in n–type GaAs bulk samples [9]. This resonance effect can be seen in more detail in Fig. 4.6. There, a series of polarized Raman spectra for many different laser energies is plotted. One can clearly see that the SPE dramatically rises in intensity if the laser frequency is tuned towards the effective bandgap and is resonantly enhanced there. On the other hand, the intersubband CDE, which are actually two modes – CDE_- and CDE_+ – due to the coupling with the LO phonon (see Sect. 3.3.3), also shows additional strong and sharp resonance maxima at higher laser energies. For laser frequencies well above the bandgap, an excitonic third–order scattering process, as for the case of quantum wells displayed in Fig. 4.4 (c), has appeared to be dominant. This mechanism was proposed by Danan et al. [7], and has been worked out theoretically by A. O. Govorov [10]. In this scattering mechanism, which is assisted by the Coulomb interaction, in the first step the incident photon creates an exciton with a hole in the valence band and an electron in a higher conduction–band state. This exciton is scattered by direct and exchange Coulomb interaction with the

[1] Within the approximations applied so far in the described theory, these excitations are *single–particle* excitations. Nevertheless, in a many–particle system all excitations must be *many–particle* excitations. Thus, a more rigorous theoretical treatment should prove that for these excitations the collective effects may be small but that they are still present.

Fermi sea into another state, where for example, as drawn in Fig. 4.4 (c), the hole is scattered into another state. By this scattering process, a collective excitation (SDE or CDE) is created in the conduction–band system. In the third step, the scattered exciton recombines. This scattering process exhibits very sharp and intense resonance profiles in a plot of the scattered intensity versus laser frequency [7]. This can be seen in Fig. 4.7. Here the integrated intensities versus laser frequency for the SPE, CDE_-, and CDE_+ are shown. The SPE exhibits a strong and broad resonance profile where presumably one of the lowest heavy–hole states of the quantum well (denoted by H_1 in Fig. 4.7) serves as an intermediate state. On the other hand, the CDE's show strong and relatively sharp resonance profiles due to the excitonic TOP. We found that this TOP is the dominant scattering mechanism for excitation of collective SDE's and CDE's in experiments on GaAs–AlGaAs samples, as will be discussed at different places in this book.

In the following we want to elucidate these scattering processes in some more detail. With (4.15), we see that the scattering cross section

$$\frac{d^2\sigma}{d\Omega\, d\omega} \propto \frac{\omega_S}{\omega_I} S(\omega) \tag{4.17}$$

is proportional to the structure factor $S(\omega)$ [11], which is defined as

$$S(\omega) = \left\langle \sum_S |\langle S|\hat{V}_{eff}|I\rangle|^2 \, \delta(E_I - E_S - \hbar\omega) \right\rangle . \tag{4.18}$$

$\hat{V}_{eff}$ is the effective operator, which describes the transition of the system from the many–particle initial state $|I\rangle$ with energy E_I to the final state $|S\rangle$ with energy E_S. Using the second–quantization technique, this effective operator of the light scattering, $\hat{V}_{eff}$, can in a many–particle system be expanded in terms of creation ($\hat{c}^\dagger$) and annihilation ($\hat{c}$) operators of single–electron states [10]

$$\hat{V}_{eff} = \sum_{\alpha,\beta} \gamma_{\alpha\beta} \hat{c}^\dagger_\beta \hat{c}_\alpha \, . \tag{4.19}$$

Here, each α and β represents a set of quantum numbers which characterize a state. If we treat for a moment the excitons as simple electron–hole pairs without interaction, which is for a quantitative analysis certainly not correct, the scattering amplitudes $\gamma_{\alpha\beta}$ can symbolically be written as [4, 10]

$$\gamma_{\alpha\beta} \sim \langle\alpha|e^{i\boldsymbol{qr}}|\beta\rangle \boldsymbol{e}_I \boldsymbol{e}_S + \frac{1}{m}\sum_{\beta'} \frac{\langle\alpha|\boldsymbol{pA}_S|\beta'\rangle\langle\beta'|\boldsymbol{pA}_I|\beta\rangle}{E_\beta - E_{\beta'} - \hbar\omega_I} + \\ + \sum_{\nu\nu'} \frac{\langle\beta|\boldsymbol{pA}_S|\nu'\rangle\langle\nu'|\hat{V}_{e-e}|\nu\rangle\langle\nu|\boldsymbol{pA}_I|\alpha\rangle}{(E_\beta - E_{\nu'} - \hbar\omega_S)\,(E_\alpha - E_\nu - \hbar\omega_I)} , \tag{4.20}$$

where with the last two terms we have written the strongest resonant terms, only. In (4.20) we have put together symbolically terms up to third–order

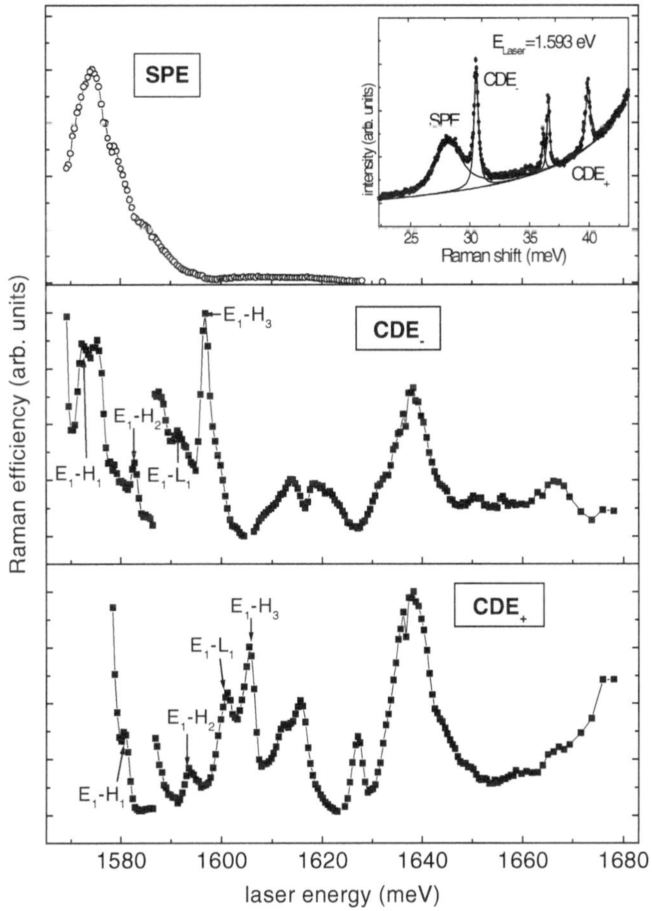

Fig. 4.7. Integrated intensities of Q2D intersubband excitations versus laser energy. The labels indicate the excitonic intermediate states. At $E_L = 1583$ meV, the y axis has been scaled down. The inset shows a typical polarized spectrum at a laser energy E_L where all excitations of interest can be observed clearly

perturbation theory. We will show below (see also [7]) that in particular the third–order term is important for scattering by charge–density fluctuations. The first term in (4.20) represents the nonresonant contributions from the $\boldsymbol{A}^2$ terms in first order perturbation theory. Here the scalar product of the polarization vectors of the incident ($\boldsymbol{e}_I$) and scattered ($\boldsymbol{e}_S$) photon shows that this nonresonant term yields only scattering by plasmons (CDE's). Spinflip processes, which are necessary for the excitation of SDE's, are possible in the resonant second and third term, only[2]. The second term describes second–order processes as schematically shown in Fig. 4.4 (a). This term leads, under

[2] This will be discussed in detail in the following two sections.

conditions of extreme resonance, to scattering by single–particle–like excitations as displayed in the uppermost panel of Fig. 4.7. The third term represents third–order resonant scattering as shown in Fig. 4.4 (c). This term causes sharp excitonic resonances in the scattering by collective excitations, as displayed in the lower panels of Fig. 4.7.

4.2.2 Scattering Cross Section: General Considerations

In this section, the scattering cross section for scattering by electron–density fluctuations – i.e., electronic excitations – via the above introduced SOP's will be investigated in detail, following a work by F. A. Blum [4, 5][3].

As discussed before, the coupling of the electromagnetic field to the carrier system yields two additional terms in the Hamiltonian,

$$H_{ep} = H''_{ep} + H'_{ep} \,, \tag{4.21}$$

with

$$H''_{ep} = \frac{e^2}{2m} \sum_j [\boldsymbol{A}(\boldsymbol{r}_j)]^2 \,, \qquad H'_{ep} = \frac{e}{2m} \sum_j [\boldsymbol{p}_j \boldsymbol{A}_j + \boldsymbol{A}_j \boldsymbol{p}_j] \,, \tag{4.22}$$

which can be treated in a perturbation approach. For our considerations here, we will use the terms up to second order, i.e., we will not consider the three–step excitonic process. This process is more involved but leads to similar results concerning scattering by spin– and charge–density fluctuations (see [10]). Single–particle–like excitations, however, appear due to the second–order processes under conditions of extreme resonance, only. This will be discussed in the following. For resonant scattering we have $\hbar\omega_I \approx E_G$, where E_G is the effective bandgap between the valence– and conduction–band states, which contribute to the scattering process. This means that in the formalism transitions across the bandgap have to be included, like it was discussed in the previous section. Furthermore, in the resonant scattering, the spin–orbit interaction plays a central role, since it allows the excitation of spin–flip transitions, which enables the scattering by spin–density fluctuations. As already mentioned, the inelastic light scattering takes place via fluctuations of the carrier density (charge–density and spin–density fluctuations). For nonresonant scattering – i.e., considering the conduction band, only – the density–fluctuation operator (or density pair operator) for a free electron gas[4], $N(\boldsymbol{q}) = \frac{1}{V} \sum_j e^{i\boldsymbol{q}\boldsymbol{r}_j}$, can in second quantization be written as

$$N(\boldsymbol{q}) = \frac{1}{V} \sum_{\boldsymbol{k}} c^\dagger_{\boldsymbol{k}+\boldsymbol{q}} c_{\boldsymbol{k}} \,. \tag{4.23}$$

[3] see also, e.g., [11, 12, 13, 14, 15, 16].

[4] The density fluctuation operator is essentially the Fourier transform of the density operator.

$c_{\boldsymbol{k}}^{\dagger}$ and $c_{\boldsymbol{k}}$ are creation and annihilation operators of single–electron states in the semiconductor, which are occupied by the free carriers. The spin of the carriers is omitted here. V is the scattering volume. If we now consider resonant scattering, an effective Hamiltonian, H_{eff}, which includes the contributions of H''_{ep} and H'_{ep} (4.22) for the coupling of the radiation field to the carrier system, can be written as [5]

$$H_{eff} = \frac{e^2}{2m} N A(\omega_I) A(\omega_S) \,. \tag{4.24}$$

$A(\omega)$ are here the amplitudes of the vector potentials of incident and scattered waves. Since in the resonant scattering process virtual transitions between states, which are outside of the state hosting the free carriers, have to be considered, N is now a generalized density pair operator:

$$N = \sum_{\alpha,\beta} \gamma_{\alpha\beta} c_{\beta}^{\dagger} c_{\alpha} \,. \tag{4.25}$$

Considering contributions up to second order, the coefficients $\gamma_{\alpha\beta}$ contain contributions of the $\boldsymbol{A}^2$ terms (intraband excitations) and the $\boldsymbol{pA}$ terms (intraband– and interband excitations). In the scattering process, an electronic excitation $|\alpha\rangle \to |\beta\rangle$ is created, where $|\alpha\rangle$ and $|\beta\rangle$ are single–particle states, which are characterized by a set of quantum numbers (band index i_α, quasimomentum $\boldsymbol{k}_\alpha$ and spin S_α). These single–particle states include the lattice–periodic Bloch part as well as the envelope function. We are here interested in the resonant contributions by interband excitations, only. Considering contributions of the $\boldsymbol{pA}$ terms in second order, we get for $\gamma_{\alpha\beta}$ in dipole approximation [4, 16]:

$$\gamma_{\alpha\beta} \approx \delta_{\boldsymbol{k}_\alpha \boldsymbol{k}_\beta} \frac{1}{m} \sum_{\beta'} \left(\frac{\langle\alpha|\, \boldsymbol{e}_I \boldsymbol{p}\, |\beta'\rangle\langle\beta'|\, \boldsymbol{p} \boldsymbol{e}_S\, |\beta\rangle}{\varepsilon_\alpha - \varepsilon_{\beta'} - \hbar\omega_I} + \frac{\langle\alpha|\, \boldsymbol{e}_S \boldsymbol{p}\, |\beta'\rangle\langle\beta'|\, \boldsymbol{p} \boldsymbol{e}_I\, |\beta\rangle}{\varepsilon_\alpha - \varepsilon_{\beta'} + \hbar\omega_I} \right) . \tag{4.26}$$

On the basis of the effective Hamiltonian, H_{eff} (4.24), for the interaction of the radiation field with the carrier system, one can derive the following expression for the scattering cross section for scattering by electron–density fluctuations [5]:

$$\frac{\partial^2 \sigma}{\partial \Omega \partial \omega} = r_0^2 \left(\frac{\omega_s}{\omega_I} \right) V^2 G(\omega) \tag{4.27}$$

with the dynamical structure factor

$$G(\omega) = \frac{1}{2\pi} \int_{-\infty}^{\infty} e^{i\omega t} < N(t) N^{\dagger}(0) > dt \,. \tag{4.28}$$

$N(t)$ is the time–dependent operator in the Heisenberg picture, which corresponds to N (4.25), and r_0 is the classical electron radius ($r_0 = e^2/4\pi\epsilon_0 mc^2$). The brackets in (4.28) mean an average over all initial states. The expression in brackets in (4.28) is the density–density correlation function of the carrier system, which is an important quantity in the description of electronic excitations. If we would consider nonresonant scattering within a one–band model, only (i.e., solely the band which is occupied by the free carriers is considered), we would get N from (4.23), and the density–density correlation function would describe pure charge–density fluctuations. Scattering by spin–density fluctuations is possible in resonant scattering, only (see Sect. 4.2.3 below). By the inclusion of resonant processes, also bands outside of the band which is occupied by the free carriers, and which are accessible by optical transitions, have to be included in the calculation. This inclusion of the band structure enters via the coefficients $\gamma_{\alpha\beta}$, which appear in the generalized density fluctuation operator (4.25), the density–density correlation function in (4.28). This means that under conditions of extreme resonance, besides charge– and spin–density fluctuations, contributions to the cross section can appear that stem from the energy–band dispersion of the involved bands. Within this level of approximation, such contributions, which are called energy–density fluctuations [17], are not subject to screening and lead to scattering by single–particle–like excitations (SPE).

So far, the discussed formalism for the determination of the scattering cross section is general and independent of special features of the electronic system. Whether or not it is possible to derive an analytical expression for the scattering cross section from (4.27) depends on the form of the generalized density pair operator N (4.25). The characteristic features of the sample, like dimensionality of the electronic system, band structure, etc., enter via the scattering amplitudes $\gamma_{\alpha\beta}$ the calculation. If it is possible to find an analytical expression for $\gamma_{\alpha\beta}$, it might be possible to derive via (4.27) and (4.28) an analytical expression for the scattering cross section, too. However, in most cases, comparison with experiment can only be made on a qualitative level and not quantitatively. One reason is that for realistic systems it is difficult to include the complex valence–band structure sufficiently. Furthermore, in the formalism described so far, e.g., Coulomb interaction is neglected in the interband transitions between valence and conduction band. In particular, the third–order excitonic scattering mechanism, discussed in the previous section, shows that Coulomb interaction can be of major importance in the intermediate steps. For the second–order processes, which are considered here in detail, Coulomb interaction is considered for the transitions within the band hosting the free carriers, via the density–density correlation function, only. First calculations of *resonant* electronic Raman scattering on semiconductor nanostructures were performed by C. Steinebach et al. [18] for quantum dots (see Sect. 5.3.4), and S. Das Sarma et al. [19] for quantum wires.

4.2.3 Scattering by Crystal Electrons: Polarization Selection Rules

The polarizations, $\boldsymbol{e}_I$ and $\boldsymbol{e}_S$, of incident and scattered radiation are included in the scattering amplitudes $\gamma_{\alpha\beta}$. In this section we will analyze the impact of the relative orientations of these polarizations. By this, we will derive the polarization selection rules for scattering by spin–density and charge–density fluctuations, and we will see that the coupling to spin–density fluctuations is possible under resonant scattering conditions, only. The considerations will be made on the basis of a simple but instructive example. As a result, we will find the well–known cross section and selection rules for the resonant scattering at the $E_0 + \Delta$ spin–orbit split–off band of n–GaAs bulk material [11].

Let us consider resonant scattering by intraband excitations of electrons, which reside in the s–type conduction band of GaAs bulk material. Let the energy of the incident laser photons, $\hbar\omega_I$, be approximately equal to the $E_0 + \Delta$ bandgap energy, i.e., the virtual interband transitions, which predominantly contribute to the scattering process, take place between the spin split–off Γ_7 valence band and the Γ_6 conduction–band states (cf. Fig. 2.4 in Sect. 2.1). In order to determine the character and polarization selection rules of the elementary electronic excitations, which we expect in the experiment, the scattering amplitude $\gamma_{\alpha\beta}$ (4.26) has to be calculated using the Γ_6 and Γ_7 band–edge Bloch functions, listed in Table 2.1 on page 14. These Bloch functions are:

$$\Gamma_6 : \begin{cases} |S\uparrow\rangle \\ |S\downarrow\rangle \,, \end{cases}$$

$$\Gamma_7 : \begin{cases} \frac{1}{\sqrt{3}}\,|(X+iY)\downarrow + Z\uparrow\rangle \; =: \; |V1\rangle \\ \frac{i}{\sqrt{3}}\,|-(X-iY)\uparrow + Z\downarrow\rangle \; =: \; |V2\rangle \,. \end{cases} \tag{4.29}$$

The initial and final states, $|\alpha\rangle$ and $|\beta\rangle$, in the scattering process are s–type functions, either with the same spin (creation of a non–spinflip excitation) or with opposite spin (creation of a spinflip excitation). The intermediate states, $|\beta'\rangle$, over which we have to sum in $\gamma_{\alpha\beta}$ [(4.26)], are in the here–considered case the Γ_7 valence–band states. Using the Kronecker products of dipole matrix elements calculated in appendix A, we receive with (4.26) and (4.29) for $\gamma_{\alpha\beta}$ for the different possible spin configurations[5]:

[5] The symbols $\uparrow$ or $\downarrow$ are used to indicate the two different spin states. For the calculation, the vector description: $|\uparrow\rangle = \binom{1}{0}$, $|\downarrow\rangle = \binom{0}{1}$ is used. Furthermore, it is assumed that we are far enough off resonance so that we can approximately replace $\varepsilon_\alpha - \varepsilon_{\beta'}$ in (4.26) by the constant $E_0+\Delta$. This means that the k dependence of $\gamma_{\alpha\beta}$ is neglected.

$$\gamma_{\uparrow\uparrow} = \frac{1}{m} \sum_{\substack{\beta'= \\ V1,V2}} \left(\frac{\langle S\uparrow | \boldsymbol{e}_i \boldsymbol{p} \, |\beta'\rangle \langle \beta'| \, \boldsymbol{p} \, \boldsymbol{e}_s \, |S\uparrow\rangle}{E_0 + \Delta - \hbar\omega_i} + \right.$$

$$\left. + \frac{\langle S\uparrow | \boldsymbol{e}_s \boldsymbol{p} \, |\beta'\rangle \langle \beta'| \, \boldsymbol{p} \, \boldsymbol{e}_i \, |S\uparrow\rangle}{E_0 + \Delta + \hbar\omega_i} \right) =$$

$$= \frac{1}{m} \frac{1}{(E_0+\Delta)^2 - (\hbar\omega_i)^2} \left[(E_0+\Delta+\hbar\omega_i) \, \boldsymbol{e}_i \, \langle S\uparrow |\boldsymbol{p}|V1\rangle\langle V1|\boldsymbol{p}|S\uparrow\rangle \, \boldsymbol{e}_s + \right.$$

$$+ (E_0+\Delta-\hbar\omega_i) \, \boldsymbol{e}_s \, \langle S\uparrow |\boldsymbol{p}|V1\rangle\langle V1|\boldsymbol{p}|S\uparrow\rangle \, \boldsymbol{e}_i \, +$$

$$+ (E_0+\Delta+\hbar\omega_i) \, \boldsymbol{e}_i \, \langle S\uparrow |\boldsymbol{p}|V2\rangle\langle V2|\boldsymbol{p}|S\uparrow\rangle \, \boldsymbol{e}_s \, +$$

$$\left. + \, (E_0+\Delta-\hbar\omega_i) \, \boldsymbol{e}_s \, \langle S\uparrow |\boldsymbol{p}|V2\rangle\langle V2|\boldsymbol{p}|S\uparrow\rangle \, \boldsymbol{e}_i \right] =$$

$$= \frac{P^2}{3m} \frac{1}{(E_0+\Delta)^2 - (\hbar\omega_i)^2} \left[(E_0+\Delta+\hbar\omega_i) \, \boldsymbol{e}_i \begin{pmatrix} 0 & 0 & 0 \\ 0 & 0 & 0 \\ 0 & 0 & 1 \end{pmatrix} \boldsymbol{e}_s + \right.$$

$$+ (E_0+\Delta-\hbar\omega_i) \, \boldsymbol{e}_s \begin{pmatrix} 0 & 0 & 0 \\ 0 & 0 & 0 \\ 0 & 0 & 1 \end{pmatrix} \boldsymbol{e}_i + (E_0+\Delta+\hbar\omega_i) \, \boldsymbol{e}_i \begin{pmatrix} 1 & i & 0 \\ -i & 1 & 0 \\ 0 & 0 & 0 \end{pmatrix} \boldsymbol{e}_s +$$

$$\left. + \, (E_0+\Delta-\hbar\omega_i) \, \boldsymbol{e}_s \begin{pmatrix} 1 & i & 0 \\ -i & 1 & 0 \\ 0 & 0 & 0 \end{pmatrix} \boldsymbol{e}_i \right] =$$

$$= \frac{2P^2}{3m} \frac{1}{(E_0+\Delta)^2 - (\hbar\omega_i)^2} \left[(E_0+\Delta) \, \boldsymbol{e}_i \begin{pmatrix} 1 & 0 & 0 \\ 0 & 1 & 0 \\ 0 & 0 & 1 \end{pmatrix} \boldsymbol{e}_s + \right.$$

$$\left. + \, \hbar\omega_i \, \boldsymbol{e}_i \begin{pmatrix} 0 & i & 0 \\ -i & 0 & 0 \\ 0 & 0 & 0 \end{pmatrix} \boldsymbol{e}_s \right] =$$

$$\gamma_{\uparrow\uparrow} = \frac{2P^2}{3m} \frac{1}{(E_0+\Delta)^2 - (\hbar\omega_i)^2} \left[(E_0+\Delta) \, \boldsymbol{e}_i \, \boldsymbol{e}_s + i \, \hbar\omega_i \, (\boldsymbol{e}_i \times \boldsymbol{e}_s)_z \right] . \tag{4.30}$$

Analogously, we get:

$$\gamma_{\downarrow\downarrow} = \gamma^*_{\uparrow\uparrow} \, , \tag{4.31}$$

$$\gamma_{\uparrow\downarrow} = \frac{2P^2}{3m} \frac{\hbar\omega_i}{(E_0+\Delta)^2 + (\hbar\omega_i)^2} \left[(\boldsymbol{e}_i \times \boldsymbol{e}_s)_y + i \, (\boldsymbol{e}_i \times \boldsymbol{e}_s)_x \right] \, , \tag{4.32}$$

$$\gamma_{\downarrow\uparrow} = -\gamma^*_{\uparrow\downarrow} \, . \tag{4.33}$$

The interband dipole matrix element, P, is defined as $P = -i\langle S|p_x|X\rangle = -i\langle S|p_y|Y\rangle = -i\langle S|p_z|Z\rangle$. Using the Pauli–spinmatrix vector $\boldsymbol{\sigma}$

$$\boldsymbol{\sigma} = \left(\begin{pmatrix} 0 & 1 \\ 1 & 0 \end{pmatrix}, \begin{pmatrix} 0 & -i \\ i & 0 \end{pmatrix}, \begin{pmatrix} 1 & 0 \\ 0 & -1 \end{pmatrix} \right) = (\sigma_x, \sigma_y, \sigma_z) \tag{4.34}$$

we can collect (4.30)–(4.33) in a single equation[6]:

$$\gamma_{\alpha\beta} = \boldsymbol{e}_i\, A\, \boldsymbol{e}_s\, \delta_{\alpha\beta} + i\, B\, (\boldsymbol{e}_i \times \boldsymbol{e}_s) \langle S_\alpha | \boldsymbol{\sigma} | S_\beta \rangle \;, \tag{4.35}$$

with

$$A = \frac{2P^2}{3m} \frac{E_0 + \Delta}{(E_0 + \Delta)^2 - (\hbar\omega_i)^2} \;, \tag{4.36}$$

$$B = \frac{2P^2}{3m} \frac{\hbar\omega_i}{(E_0 + \Delta)^2 - (\hbar\omega_i)^2} \;. \tag{4.37}$$

S_α, S_β shall symbolize the different spin states. With (4.35) we see that the inelastic light scattering by electron–density fluctuations consists of two parts: One is symmetric in the polarizations of incident and scattered light, and the other is asymmetric in the polarizations. The symmetric part stems from the scattering by charge–density fluctuations, while the asymmetric one is from scattering by spin–density fluctuations. This can easily be understood in the following way: Inserting the symmetric part of (4.35) into the generalized density pair operator (4.25), delivers for the interaction Hamiltonian (4.24):

$$H'_{eff} = \frac{e^2}{2m} (\boldsymbol{e}_i\, A\, \boldsymbol{e}_s) \left(\frac{N_\uparrow(q) + N_\downarrow(q)}{2} \right) A(\omega_i) A(\omega_s) \;. \tag{4.38}$$

$N(q)$ here is the density pair operator (4.23) for a given spin direction. Equation (4.38) shows that the light wave couples via the symmetric part of (4.35) to the charge–density fluctuations, $e(N_\uparrow(q) + N_\downarrow(q))$, of the system. Inserting analogously the asymmetric part of (4.35) into (4.25), and using the z component, σ_z, of the Pauli–spinmatrix vector (4.34) in (4.35) yields[7]:

$$H''_{eff} = i \frac{e^2}{2m} (\boldsymbol{e}_i \times \boldsymbol{e}_s)\, B\, \boldsymbol{e}_z \left(\frac{N_\uparrow(q) - N_\downarrow(q)}{2} \right) A(\omega_i) A(\omega_s) \;. \tag{4.39}$$

One recognizes with (4.39) that – even though σ_z does not flip the spin of the electron – the light wave couples to the spin–density fluctuations, $N_\uparrow(q) - N_\downarrow(q)$. Only employing σ_y in (4.35) flips the spin of the electron. The resulting Hamiltonian, H''_{eff}, would be equivalent to (4.39). The only differences that would occur in (4.39) due to the choice of different σ_μ, are the relative orientations of $\boldsymbol{e}_i$ and $\boldsymbol{e}_s$ with respect to the quantization direction of the spin. The band–edge Bloch functions (4.29) are chosen such that the z direction is the direction of spin quantization. Without external magnetic field, however, the quantization direction of the spin can be arbitrarily chosen. Accepting this, one can in (4.39) no longer differentiate between spinflip and

[6] This is the same equation, that appears in [11] for resonant scattering at the $E_0 + \Delta$ bandgap.

[7] This means that the spin configuration $(S_\alpha, S_\beta) = (\uparrow, \uparrow)$ was chosen, i.e., spinflip excitations are <u>not</u> considered. With (4.34) it is: $\langle \uparrow | \boldsymbol{\sigma} | \uparrow \rangle = (0, 0, 1) = \boldsymbol{e}_z$.

nonspinflip excitations. Hence it follows that without external magnetic field, spinflip and spin–density excitations are identical.

Furthermore, it shall be noted that the asymmetric part of (4.35) – i.e., the scattering by spin–density fluctuations – crucially depends on the spin–orbit interaction. Equation (4.29) shows that the Γ_7 band–edge Bloch functions are no eigenfunctions of the spin, since they are a superposition of different spin functions, which is a consequence of spin–orbit interaction. This is the origin of the asymmetric term in (4.35). A further important precondition for the occurrence of spin–density excitations is resonant scattering: The Γ_7 states are included via virtual interband transitions into the scattering process. One can see with (4.35) and (4.37) that with $\omega_I \to 0$ the scattering amplitude, B, for scattering by spin–density fluctuations goes to zero, too. On the other hand, for $\hbar\omega_I = E_0 + \Delta$, B is going to be maximal. Finally, considering the resonance at the E_0 band gap delivers with the heavy– and light–hole states of the Γ_8 band edge (see Table 1, on page 14) as intermediate states in the scattering process similar expressions for the scattering amplitude (4.35) (see also [11]).

With the discussed simple example, we can deduce important general statements about polarization selection rules in semiconductor nanostructures with Zincblende bulk–lattice structure: As long as the effective–mass approximation is a valid approximation to describe the band structure of the nanostructures, the polarization selection rules, namely spin–density excitations occur in depolarized scattering geometry, i.e.,

$$\boldsymbol{e}_i \perp \boldsymbol{e}_s \, , \tag{4.40}$$

and, charge–density excitations are allowed in polarized configuration, i.e.,

$$\boldsymbol{e}_i \parallel \boldsymbol{e}_s \tag{4.41}$$

hold for nonresonant conditions. Nonresonant conditions means that the explicit $\boldsymbol{k}$ dependence of the involved states can be neglected, as we have done above. In the case of extreme resonance, this $\boldsymbol{k}$ dependence has to be included, explicitly [17]. This leads, as already mentioned, to the scattering by so called energy–density fluctuations (single–particle–like excitations): The $\boldsymbol{k}$ dependent band structure enters via the generalized pair operator (4.25) the interaction Hamiltonian (4.24). For extreme resonance conditions, this $\boldsymbol{k}$ dependence can not be neglected. This was experimentally recognized first in experiments of A. Pinczuk et al. on n–GaAs bulk material [9], and theoretically explained by M. V. Klein [17]. In later chapters of this book we will see that also in lower–dimensional structures, like quantum wires and dots, extreme resonance leads to scattering by single–particle–like excitations.

We have seen above that the basic polarization selection rules for nonresonant conditions originate from the *Bloch parts* of the wave functions. This means that, first of all, they are independent of the dimensionality of the electronic system. The dimensionality enters via the envelope functions. As mentioned before, the single–particle states, which occur in the scattering amplitude (4.35), consist of the Bloch part *and* the envelope wave functions, χ_i, as introduced in Sect. 2.4. For inelastic light scattering on semiconductor nanostructures this means that, on top of the general polarization selection rules for scattering by spin–density and charge–density excitations, these envelope wave functions can give rise to parity selection rules.

4.2.4 Parity Selection Rules in Nanostructures

Investigating the scattering amplitude for second–order processes, $\gamma_{\alpha\beta}$ (4.35), we see that, besides dipole matrix elements between Bloch states, we have to consider products of overlap integrals of the involved envelope functions[8]. This is of importance for structures with dimensionality lower than 3D, since there the envelope functions determine the quantization effects, and are generally aperiodic functions. If we, e.g., consider a resonant scattering process, as displayed in Fig. 4.4 on page 63 for scattering on intersubband excitations, $j \to j+1$, of a Q2D systems, we get for the envelope part of the scattering amplitude

$$\gamma_{jj+1} \propto \int \chi_i^{vb}(z)\chi_{j+1}^{cb}(z)dz \int \chi_i^{vb}(z)\chi_j^{cb}(z)dz \; . \tag{4.42}$$

Here, j labels the conduction–band subbands, and, i the valence–band states. With (4.42) we find a very general result concerning parity selection rules in symmetric structures. For a symmetric square–well potential of infinite height, e.g., the second term in (4.42) is nonzero for $i = j$, only. However, then the first term is vanishing. We would get a weakly–allowed transition for $j \to j+2$, and a finite potential height, only. Therefore, for a strictly symmetric well, excitations $j \to j+1$, as displayed in Fig. 4.5 on page 64 are parity forbidden and should not be visible in experiment. This is a further important reason why for inelastic light scattering experiments *asymmetric* wells, like displayed in Fig. 2.9 on page 27 are commonly used, since for an asymmetric potential the parity selection rules are softened. Our finding here concerning parity selection rules in a symmetric potential matches the exclusion principle of far–infrared and Raman spectroscopy on vibrations of molecules: There, for symmetric molecules, Raman–allowed modes are symmetric vibrations, i.e., they have even parity, and far–infrared allowed modes have odd parity. This can, on general grounds, be traced back to the one and two–photon nature of the absorption and Raman process, respectively. Since the Raman process is a two–photon process, the allowed excitations have even parity. We will find this general rules again in the next two chapters, where experiments on quantum dots and quantum wires will be presented. At this point it should also be noted that, additionally to these intrinsic selection

[8] Under the assumption that the dipole operator acts on the Bloch part of the total wave function, only.

rules, in nanostructured samples with a periodic pattern, optical near–field effects can play a significant role. Due to sample geometries, the polarizations can be different in the near and in the far fields. Such effects were recognized first in experiments on quantum wires by C. Dahl et al. [20].

In conclusion, we can say that the results concerning polarization selection rules, derived in the previous section, can be generalized to systems with lower dimensionality. For those systems we would get similar results for the scattering amplitude (4.35). However, the coefficients A (4.36), and B (4.37) would get a more complex form, since overlap integrals of envelope wave functions have to be considered, which in addition lead to parity selection rules.

4.2.5 Intrasubband Excitations, Grating Coupler–Assisted Scattering

After considering the scattering processes in detail, this chapter is finalized by two subsections about experiments on Q2D *intrasubband* excitations. In Sect. 4.2.1 intersubband excitations of Q2D quantum wells were presented and discussed. We saw that a particular strength of the inelastic light scattering technique is the possibility to transfer a finite quasimomentum q to the excitation. This allows one to excite intraband excitations in Q2D systems. The energy of the intraband CDE depends on the square root of q (cf. Sect. 3.8). Figure 4.8 displays polarized spectra of a GaAs–AlGaAs single quantum well with a carrier density of $N_s = 4 \times 10^{11}$ cm^{-2} for different

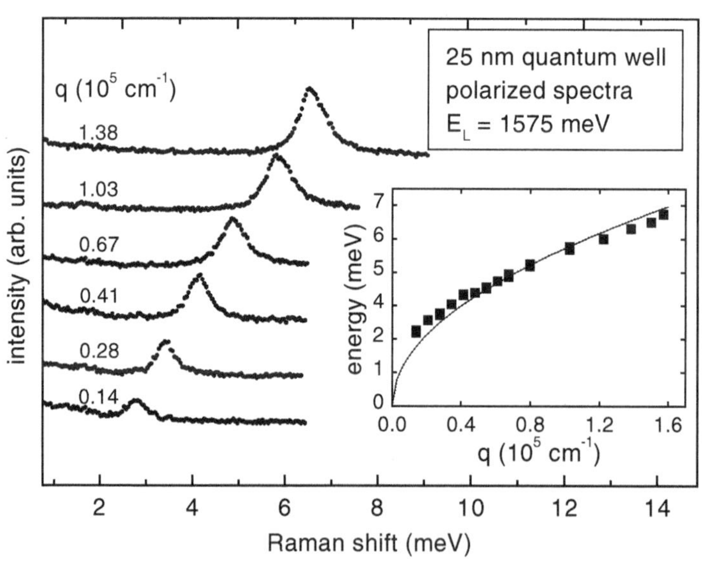

Fig. 4.8. Polarized spectra of a GaAs–AlGaAs quantum–well sample for different wave–vector transfer q. The inset shows the measured excitation energies versus q

wave–vector transfer q parallel to the quantum–well plane. The inset shows the measured mode positions versus wave–vector transfer q. One can see that the square–root dispersion is reasonably–well fulfilled. Deviations can be due to the fact that in the real sample the dielectric constant is not homogeneous over the whole space rather it is in the simplest case a step–like function in growth direction with a step at the sample surface. In our experiments on many different Q2D samples we found that the relative strengths of *intersubband* CDE, SDE, and SPE do not significantly depend on the doping levels of the samples, i.e., the 2D carrier densities. On the contrary, there was a significantly different behavior found for the *intrasubband* excitations. The spectra of the intrasubband CDE, displayed in Fig. 4.8, were recorded on a sample with relatively low carrier concentration. In such samples, typically, the intrasubband CDE occurred as a relatively sharp line, whereas the SDE was only very weak or even completely undetectable, so were the intrasubband SPE's in both polarizations. On the other hand, in samples with higher N_s of about $N_s > 5 \times 10^{11}$ cm^{-2}, intrasubband SPE's were relatively strong but the intrasubband CDE weak and typically very broad. A possible explanation for the damping of the CDE in strongly–doped samples could be a parallel conductance in the doping layer in the barrier. Due to stronger doping, also in the AlGaAs barrier layer free carriers can exist. These carriers have a very low mobility, since they reside in a region with strong disorder and Coulomb scattering due to the positively–charged donors. Macroscopically, in the intrasubband excitations, the carriers oscillate *parallel* to the plane of the quantum well (cf. Sect. 3.8). Via Coulomb interaction, both layers can couple and then the mode width is governed by the sheet with the lower mobility, i.e., the barrier layer. This mechanism would, on the other hand, not be effective for *intersubband* excitations, since in the barrier layer we should not expect intersubband transitions with nearly the same energy as in the quantum well, so that no effective coupling between both layers should be expected. This explains why – independent of the doping level – the intersubband excitations appear undamped, as sharp lines, whereas the damping of intrasubband excitations can depend on the doping level.

Figure 4.9a displays polarized spectra of intrasubband SPE of a sample with a rather large density of $N_s = 6.5 \times 10^{11}$ cm^{-2}, where the SPE's are well pronounced. In Sect. 3.8 we have seen that in Q2D systems the intrasubband SDE lies completely within the intrasubband SPE continuum and should therefore be strongly Landau damped. For comparison, there is one depolarized spectrum shown in Fig. 4.9. One can see that the maximum of the signal is slightly shifted to lower energies, compared to the polarized spectrum. This shift can be due to the Landau–damped intrasubband SDE, whose energy should be slightly below the border of the intrasubband SPE continuum. Figure 4.9b displays the mode positions of both polarizations versus q. As expected for intrasubband SPE, the slope is in good approximation linear with q (cf. Sect. 3.8). From the linear slope the carrier density N_s can

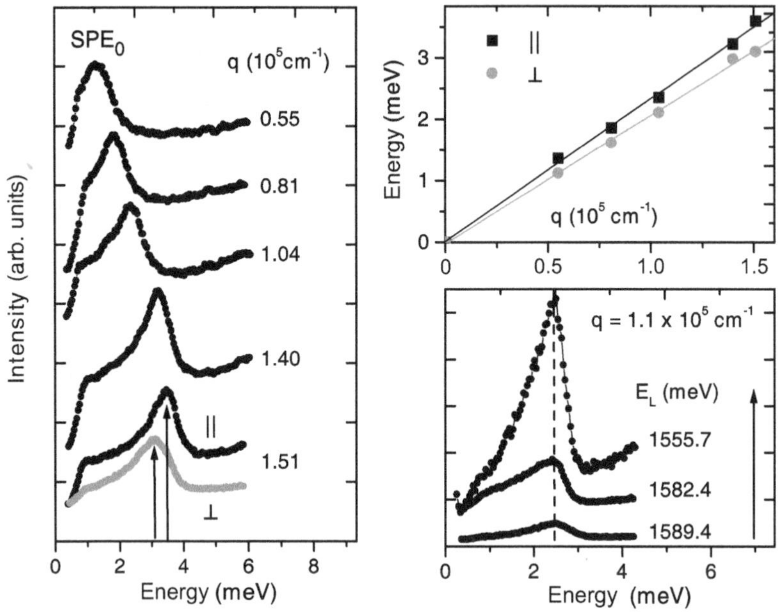

Fig. 4.9. (**a**) Polarized and depolarized spectra of intrasubband excitations in a GaAs–AlGaAs quantum–well sample with $N_s = 6.5 \times 10^{11}$ cm^{-2} for different wave–vector transfer q. (**b**) Mode positions versus q. (**c**) Polarized spectra for different laser energies E_L

be deduced via $E \approx \frac{\hbar^2}{2m^*} q k_F = \frac{\hbar^2}{2m^*} q\sqrt{2\pi N_s}$. Figure 4.9c demonstrates the resonant enhancement of the SPE if the laser energy is tuned towards the bandgap energy.

T. Zettler et al. used for the first time a so called grating coupler to enlarge the wave–vector transfer into the sample in inelastic light scattering experiments [21, 22]. A grating coupler is a dielectric grating, which is superimposed on the sample surface. Via this dielectric grating, multiples of reciprocal grating vectors, $\boldsymbol{g}$, where $|\boldsymbol{g}| = 2\pi/d$ if d is the period of the grating, can be transferred to the system. The total quasimomentum, q_{total}, which is transferred in the experiment, depends on the relative direction of the momentum $\boldsymbol{q}$ of the incoming light with respect to the direction of the grids in the grating. If $\boldsymbol{q}$ is perpendicular or parallel to the direction of the grids, the total transfer is given by

$$q_{\text{total}} = q \pm ng \tag{4.43}$$

or

$$q_{\text{total}} = \sqrt{q^2 + n^2 g^2}\ , \tag{4.44}$$

respectively, where $n = 0, 1, 2, \ldots$. In Fig. 4.10 the results of experiments on Q2D samples with two different grating couplers with periods of $d = 500$ nm and $d = 600$ nm are summarized. In that case the grating coupler consisted of an array of Titanium stripes on the sample surface. In each spectrum several peaks could be observed due to the transfer of multiples of the reciprocal grating vector. An exemplary spectrum is displayed in the left panel. In experiments without grating coupler, the maximally transferred quasimomentum is about 1.5×10^5 cm^{-1}. One can see in Fig. 4.10 that the grating coupler–assisted experiments allows one to follow the dispersion of the intrasubband CDE to much larger momenta. It can also be seen that for the Q2D quantum well, which was investigated in this experiment, the $\sqrt{q}$ relation of the intrasubband plasmon is fairly well fulfilled.

4.2.6 Multiple Cyclotron Resonance Excitations in Quantum Wells

It is also of interest, in which way a magnetic field affects the behavior of electronic excitations under different resonance conditions. In a Q2D system, for instance, the application of a perpendicular magnetic field, B, induces a zero–dimensional density of states. The single–particle–like intrasubband excitations in this case are cyclotron–resonance excitations, i.e., transitions between neighboring Landau levels. We want to focus here in particular on

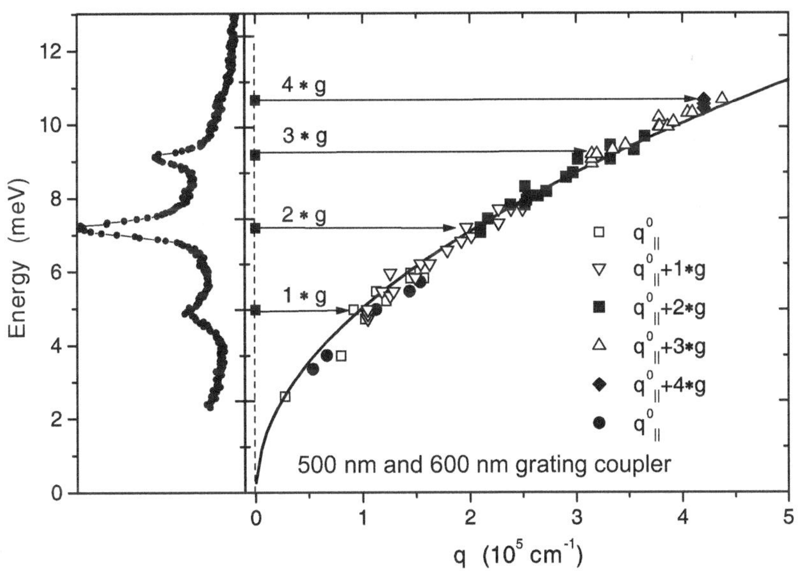

Fig. 4.10. Measured mode positions of the intrasubband CDE in samples of the same wafer with two different grating couplers. The *left panel* shows a spectrum where different peaks due to transfer of different multiples of the reciprocal grating vector can be seen

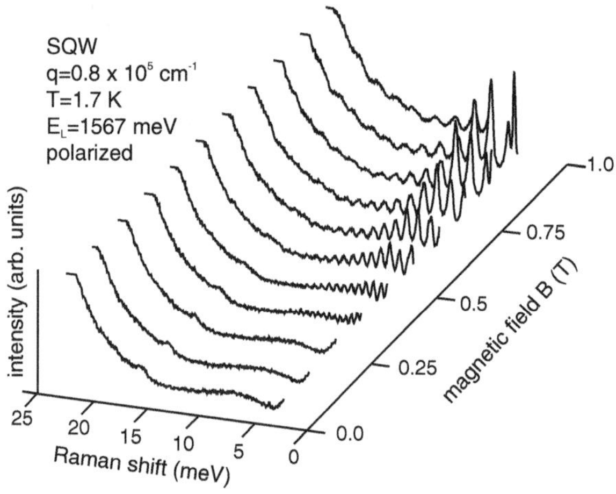

Fig. 4.11. Polarized inelastic light scattering spectra of a quantum–well sample for small magnetic fields $B = 0 - 1$ T [Reprinted with permission from [23]. Copyright (1997) by the American Physical Society]

what happens under conditions of extreme resonance, i.e., where at $B = 0$ SPE dominate the spectra. Here, in experiments on quantum wells, unexpected multiple cyclotron resonance excitations, $n\hbar\omega_c$, were found [23]. These multiple excitations were explained to stem from a higher–order cascade–like scattering process, where each step in the scattering process is very similar to the TOP discussed in Sect. 4.2.1. In this last section of the introductory part, this new scattering process shall be introduced.

Figure 4.11 shows polarized Raman spectra of a quantum–well sample in dependence on magnetic field [23]. At $B = 0$ T, no excitation in the displayed energy range is observed. For $B > 0.3$ T, several equally spaced peaks occur, where their energetic positions differ by exactly $\hbar\omega_c$. For Raman shifts >22 meV, the cyclotron excitations are obscured by the much stronger 2D intersubband excitations and luminescence. It has been found that the cyclotron resonance excitations occur in the depolarized as well as in the polarized scattering configurations, where the observed spectra do not differ significantly in both polarizations. This single–particle character is also emphasized by the observation that the excitations show strong resonant behavior in the vicinity of the E_0 gap. If the laser energy is well above the effective bandgap of the quantum well, only one single cyclotron resonance peak at $\hbar\omega_c$ is observed. The multiple resonances occur under conditions of extreme resonance in a range of laser energies, where in the quantum–well samples as well as in laterally structured samples, at $B = 0$ T, single–particle–like excitations can be observed [2].

The magnetic–field dispersion of the cyclotron resonances is shown in Fig. 4.12. The solid dots in the inset are the observed mode positions in the

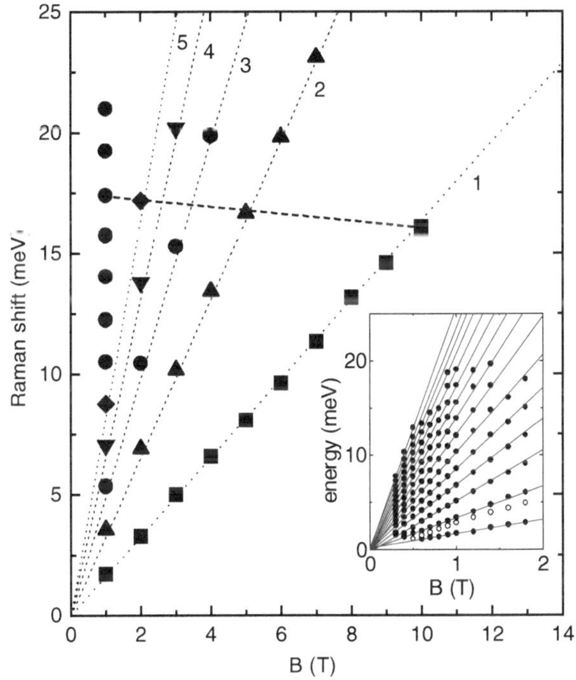

Fig. 4.12. Magnetic–field dispersion of the multiple cyclotron resonance excitations. The numbers n label the number of excited cyclotron quanta [Reprinted with permission from [23]. Copyright (1997) by the American Physical Society]

range $B = 0 - 2$ T. The dotted lines in Fig. 4.12 represent the calculated cyclotron resonance dispersions $n\hbar\omega_c = n\hbar\frac{eB}{m^*}$, where the effective mass is m^* $=0.071\ m_e$, and $n = 1, 2, 3, \ldots$. The deviation from the linear dispersion relation in the range of 0 to 10 T is about 9 % and is due to the nonparabolicity of the GaAs bandstructure. It is emphasized that, for a fixed magnetic field, the energy spacing between the multiple resonances is the same to an accuracy of 1.5 %. For clarity, the dashed line in Fig. 4.12 connects the tenth point at 1 T with the first point at 10 T. If the multiple excitations were higher harmonics, i.e., inter–Landau–level transitions with change in Landau–level quantum number $\Delta l > 1$, the dashed line should be horizontal. This leads to the conclusion that the multiple cyclotron resonances all arise from Landau–level transitions with $\Delta l = 1$. On the other hand, if higher–order Raman scattering would be the reason for the observed peaks, the intensity of the $n = 2$ excitation should be at least one order of magnitude less than the intensity of the first cyclotron resonance peak. Obviously, this is not the case. It has been suggested [23] that the multiple cyclotron resonances are not created simultaneously, but in a cascade process. In one step of this process, the photo–excited electron–hole pair is scattered via an interaction with the Fermi sea by creating a cyclotron–resonance excitation. The number of cyclotron resonances created in one cascade seems to be arbitrary. Energy conservation is provided by the almost continuous distribution of hole states in the valence band and thus all the scattering processes can take place via real transitions. One necessity for such cascades to occur is that the lifetime for creation of a cyclotron resonance excitation is much shorter than the lifetime for radiative recombination. Figures 4.13 (a), (b), and (c) show a schematic picture of the proposed scattering process: In the first step (1) an electron–hole pair or exciton is created by absorption of the incident photon. In the second step (2), the hole is scattered into another state due to Coulomb interaction with the Fermi sea, where during the scattering process a cyclotron resonance excitation in the conduction band is created. We believe that these two steps of the scattering process are very similar to the excitonic TOP discussed in Sect. 4.2.1. In the third step (3) the electron recombines with the scattered hole. The important point is that due to the almost continuous distribution of real states in the valence band with finite magnetic field, the hole may be scattered n times, where all intermediate states in the scattering or relaxation process may be real, before it recombines. During these n scattering processes, n cyclotron resonance excitations can be created in the conduction band by energy conservation (Fig. 4.13 (b) and (c)). In principle, the hole can resonantly relax until it reaches the top of the valence band.

As mentioned above, at laser energies well above the bandgap, we observe only one cyclotron resonance peak. At these laser energies, also the

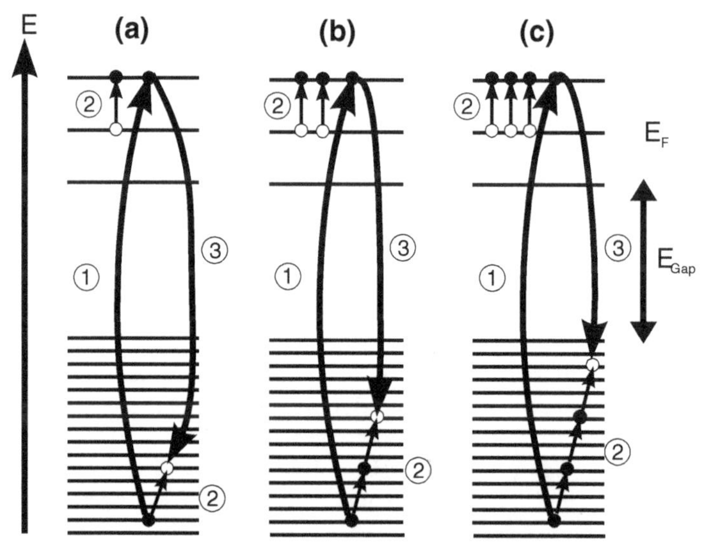

Fig. 4.13. Sketch of the proposed cascade–like scattering process, where (**a**) one, (**b**) two, or (**c**) three cyclotron–resonance excitations are created [Reprinted with permission from [23]. Copyright (1997) by the American Physical Society]

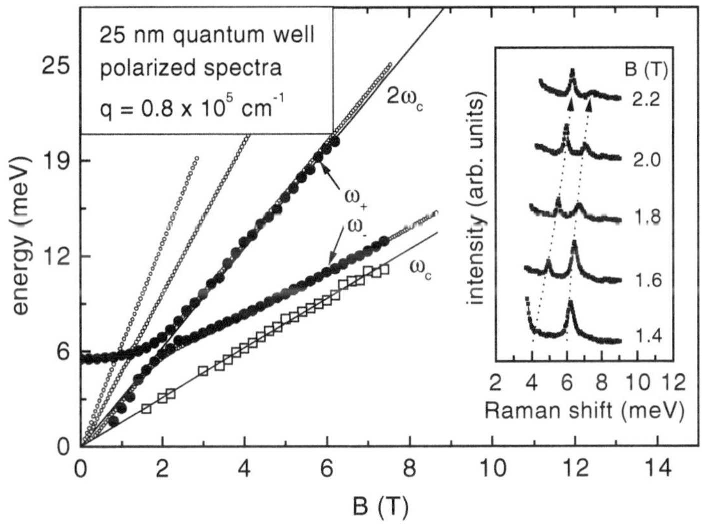

Fig. 4.14. Magnetic–field dispersion of the intrasubband magnetoplasmon mode in a quantum–well sample. The inset shows polarized spectra in the range where the anticrossing occurs

collective intrasubband plasmon can be observed. Figure 4.14 shows the measured magnetic–field dispersion of the intrasubband CDE and the cyclotron–resonance excitation for a quantum–well sample with a carrier density of $N_s = 4 \times 10^{11}$ cm^{-2} (same sample as in Fig. 4.8) and fixed wave–vector transfer q. At $2\omega_c$, a strong anticrossing due to the coupling with Bernstein modes [25, 26] is observed. In Chap. 6 we will discuss in detail similar internal interaction effects of electrons in Q1D quantum wires [27, 28].

References

1. see, e.g., W. Hayes, R. Loudon: *Scattering of Light by Crystals* (John Wiley and sons, New York, 1978)
2. C. Schüller, G. Biese, K. Keller, C. Steinebach, D. Heitmann, P. Grambow, and K. Eberl: Phys. Rev. B **54**, R17304 (1996)
3. Christian Schüller: in Festkörperprobleme / Advances in Solid State Physics, Ed. B. Kramer (Vieweg Braunschweig Wiesbaden 1999) Vol. 38, p. 167
4. F. A. Blum: Phys. Rev. B **1**, 1125 (1970)
5. G. Abstreiter, M. Cardona, A. Pinczuk in: *Light Scattering in Solids IV*, Topics in Applied Physics Vol. 54 (Springer, Berlin, Heidelberg, New York, Tokyo) 1984, Eds. M. Cardona and G. Güntherodt, p. 5
6. E. Burstein, A. Pinczuk, D. L. Mills: Surf. Sci. **98**, 451 (1980)
7. G. Danan, A. Pinczuk, J. P. Valladares, L. N. Pfeiffer, K. W. West, and C. W. Tu: Phys. Rev. B **39**, 5512 (1989)

8. A. Pinczuk, S. Schmitt–Rink, G. Danan, J. P. Valladares, L. N. Pfeiffer, and K. W. West: Phys. Rev. Lett. **63**, 1633 (1989)
9. A. Pinczuk, L. Brillson, and E. Burstein: Phys. Rev. Lett. **27**, 317 (1971)
10. A. O. Govorov: J. Phys. Condens. Matter **9**, 4681 (1997)
11. D. Hamilton, A. L. McWhorter, Light Scattering Spectra of Solids, Ed. G. B. Wright (Springer, New York, 1969) p. 309
12. S. S. Jha: Phys. Rev. **150**, 150 (1966)
13. S. S. Jha: Phys. Rev. **150**, 413 (1966)
14. S. S. Jha: Phys. Rev. **182**, 815 (1969)
15. S. S. Jha: Phys. Rev. **179**, 764 (1969)
16. S. S. Jha: Nuovo Cimento **B** LXIII, 331 (1969)
17. M. V. Klein in: *Light Scattering in Solids*, Topics in Applied Physics Vol. 8 (Springer Berlin Heidelberg New York) 1975, Ed. M. Cardona, p. 147
18. C. Steinebach, C. Schüller, and D. Heitmann: Phys. Rev. B **59**, 10240 (1999)
19. S. Das Sarma and Daw-Wei Wang: Phys. Rev. Lett. **83**, 816 (1999)
20. C. Dahl, B. Jusserand, and B. Etienne: Phys. Rev. B **51**, 17211 (1995)
21. T. Zettler,C. Peters, and J. P. Kotthaus: Phys. Rev. B **39**, 3931 (1989)
22. T. Zettler, C. Peters, and J. P. Kotthaus: Surf. Sci. **229**, 388 (1990)
23. C. Schüller, R. Krahne, G. Biese, C. Steinebach, E. Ulrichs, D. Heitmann, and K. Eberl: Phys. Rev. B **56**, 1037 (1997)
24. I. B. Bernstein: Phys. Rev. **109**, 10 (1958)
25. E. Batke, D. Heitmann, J. P. Kotthaus, and K. Ploog: Phys. Rev. Lett. **54**, 2367 (1985)
26. C. K. N. Patel and R. E. Slusher: Phys. Rev. Lett. **21**, 1563 (1968)
27. C. Steinebach, R. Krahne, G. Biese, C. Schüller, D. Heitmann, and K. Eberl: Phys. Rev. B **54**, R14281 (1996)
28. E. Ulrichs, G. Biese, C. Steinebach, C. Schüller, and D. Heitmann: Phys. Rev. B **56**, R12760 (1997)

Part II

Recent Advances

5 Quantum Dots: Spectroscopy of Artificial Atoms

5.1 Introduction

Semiconductor quantum dots are fascinating objects, since, in some respect, they can be regarded as artificial atoms [1]. Figure 5.1 shows a very schematic comparison of a real three–dimensional atom and a disc–shaped quantum dot. The structure of real atoms is three–dimensional, while most of the artificial quantum dots can be regarded as large Q2D atoms, since the lateral dimensions are in most cases much larger than the vertical extension. Of course, a crucial difference between the two systems is the shape of the confining potentials, which, for real atoms is essentially the Coulomb potential of the nucleus, and, for quantum–dot atoms in some approximation a two–dimensional parabolic potential. In Fig. 5.2, the resulting energy–level structures for a Hydrogen atom and a quantum dot, containing a single electron, are displayed schematically. Due to the reduced symmetry of the disc–shaped dot, the degeneracy of the energy levels is smaller than in the

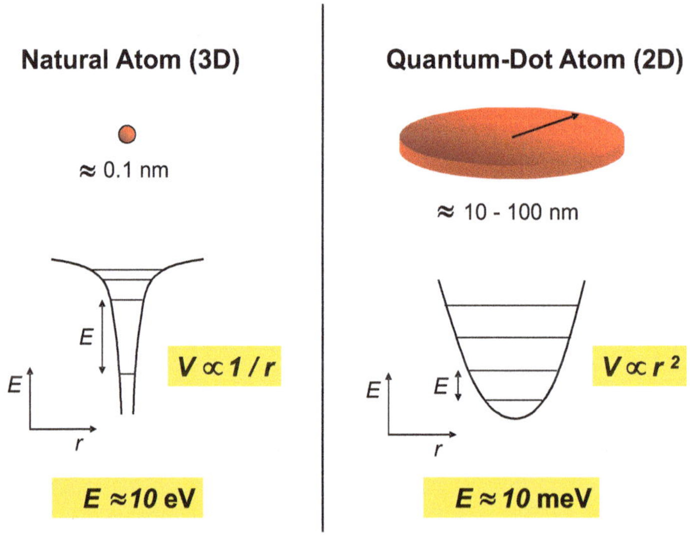

Fig. 5.1. Schematic comparison of a real atom (*left*) and a quantum–dot atom (*right*) represented by a disc–shaped two–dimensional quantum dot

Christian Schüller: *Inelastic Light Scattering of Semiconductor Nanostructures*
STMP **219**, 87–119 (2006)
DOI 10.1007/3-540-36526-5_5

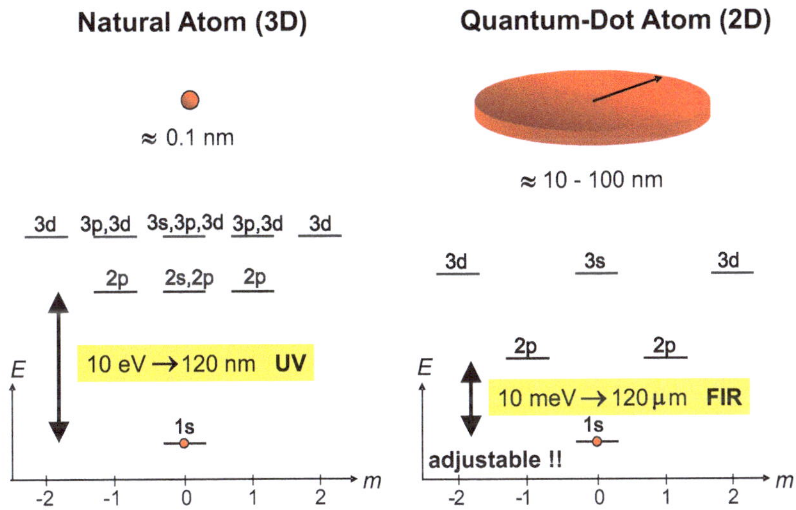

Fig. 5.2. Schematic comparison of the energy–level structure of a Hydrogen atom (*left*) with a quantum–dot Hydrogen atom, i.e., a quantum dot containing a single electron

three–dimensional case. The most striking difference, however, is that in a semiconductor quantum dot, both, the level structure and the order number, i.e., the number of electrons in the quasiatom, is tunable to some extend. Furthermore, the quantization energy is in a quantum dot of the same order of magnitude as the electron–electron Coulomb interaction, in contrast to real atoms, where the quantization is typically much larger. From this one can expect in quantum dots an interesting interplay between quantization– and Coulomb interaction–induced effects, as will be discussed later.

In this chapter we will see that in these well–defined quantum–mechanical structures, e.g., parity selection rules for inelastic light scattering and the exclusion principle between far–infrared (FIR) and inelastic light scattering spectroscopy, as generally discussed in the first part of this book, can be nicely studied. Historically, electronic excitations in lithographically–defined quantum dots were studied first by FIR transmission spectroscopy, starting in 1988 [2, 3, 4, 5, 6, 7, 8]. In the following years, also resonant Raman scattering, i.e., inelastic light scattering, was successfully applied to study the spectrum of elementary excitations of these systems [9, 10, 11, 12, 13]. As introduced in the first part, for quantum wells it is well known that besides collective spin–density (SDE) and charge–density excitations (CDE) one can observe nearly unrenormalized excitations – so called single–particle excitations (SPE) – in inelastic light scattering experiments. Both, SDE and CDE are collective excitations, where SDE are affected by exchange interaction and CDE by the full Coulomb interaction of the electrons. However, the origin of the SPE's, which seem to be unaffected by the particle–particle interaction, has posed a puzzle. In an experimental work, it was shown that the SPE's can also be observed in quantum wires and quantum dots under conditions of extreme resonance, when the laser energy is close to the fundamental bandgap of the structures [11]. Then, the SPE's are created in a resonant density–fluctuation scattering process [14, 15], whereas the collective SDE and CDE are due to an excitonic third–order scattering process [11, 16, 17] (see discussion in Chap. 4 and Sect. 5.3). In this chapter we will see that – at least for quantum dots – the SPE's are actually *collective* excitations: SDE's and CDE's. However, the many–particle interaction effects partly cancel under specific conditions so that the energies are close to single–particle energies of a noninteracting system.

Most of the available theories of nonresonant Raman scattering on quantum dots accurately describe the energetic positions of the collective excitations as well as the wave–vector and magnetic–field dependence of the CDE and SDE [18, 19, 20, 21]. However, they fail in predicting the experimentally observed relative strengths of the different modes. Furthermore, the occurrence of SPE cannot be explained within these theories. It has been known for a long time that for a correct treatment of the resonant scattering cross section, valence–band states play a crucial role [22, 23]. More recent theoretical papers on quantum wires [15, 24, 25] and quantum dots [26] showed that the inclusion of the valence–band states indeed changes the intensities of the excitations significantly. This is one of the topics, which will be addressed in detail in Sect. 5.3.

The lithographically–defined quantum dots – mostly prepared on GaAs–AlGaAs heterostructures or quantum wells – have lateral dimensions on the order of 100 nm and relatively large electron numbers. Thus, they can be regarded as mesoscopic dots – kind of in between a two–dimensional and a zero–dimensional system. During the late 1990's, self–assembled InAs quantum dots (SAQD) came up and have proven to be highly interesting quantum structures both from a technological as well as from a fundamental physics point of view (see Sect. 2.5.3). They exhibit relatively large quantization energies in the range of about 50 meV, in contrast to the quantization energies in lithographically–defined structures, which are typically about one order of magnitude smaller. In most experiments reported so far, SAQD have been investigated by optical spectroscopy, in particular photoluminescence (PL). Nowadays, PL experiments on single dots are well established, which overcome the inhomogeneously broadened linewidths in typical ensemble measurements [27]. It has also been demonstrated that it is possible to charge SAQD with single electrons [28] by application of external gate structures. So far, there are only two reports in literature about inelastic light scattering experiments on electronic excitations in InGaAs SAQD [29, 30]. Later in this chapter, inelastic light scattering experiments on collective CDE's in InAs SAQD with tunable electron numbers, N, will be described. In these experiments, N could be controlled between $N = 1 \ldots 6$ [30].

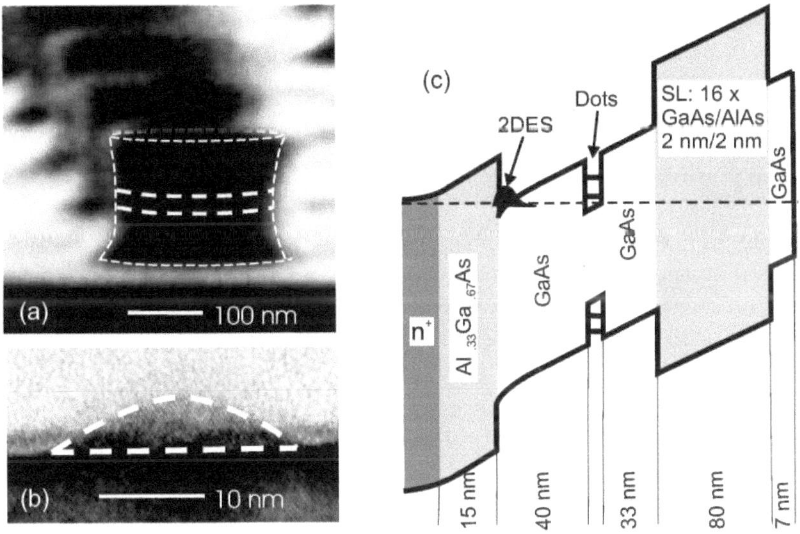

Fig. 5.3. (**a**) Scanning electron micrograph of deep-etched GaAs-AlGaAs quantum dots. The *white dashed lines* mark one of the dots. The *long dashes* indicate the active electron layer. (**b**) Transmission electron micrograph of an InAs quantum dot. The *white dashed lines* mark the borders of the InAs island [with permission of S. Mendach]. (**c**) Schematic band structure of an InAs SAQD sample with a two-dimensional electron system (2DES) as a back contact

5.2 Semiconductor Quantum Dots

5.2.1 Preparation of Quantum Dots

The vast majority of inelastic light scattering experiments on electronic excitations in quantum dots have been performed on modulation–doped GaAs–AlGaAs structures. Figure 1a shows a scanning electron micrograph of lithographically–defined, deep–etched quantum dots (see also Sect. 2.5). Typical lateral sizes of these structures are on the order of several hundred nanometers. In many cases, Q2D electron systems, realized in modulation–doped single quantum wells (cf. Sects. 2.3 and 2.4), are used as a starting material. In Fig. 5.3a, the location of such a 25–nm–wide quantum well is indicated by thick dashed lines. By a reactive–ion etching process (Sect. 2.5.2), the pillar–shaped quantum dots were defined. Typically, such deep–etched structures contain electron numbers on the order of several hundreds. An alternative and very effective way to produce quasi zero–dimensional semiconductor structures is the so-called self–assembled growth of InAs quantum dots (see Sect. 2.5.3). Here, the lattice mismatch between GaAs, which is used as the starting material, and InAs is exploited. The minimization of strain energy leads to the formation of small InAs islands, if InAs is grown on a GaAs surface. In contrast to the etched structures, the typical lateral sizes of these islands are on the order of only a few tens of nanometers. Therefore, they exhibit much larger quantization energies. Figure 5.3b displays a transmission electron micrograph of an InAs quantum dot, embedded in GaAs. The interfaces between the InAs dot and the surrounding GaAs are indicated by a white dashed line. It has been shown that electrons can be filled into these quantum boxes either by modulation doping [29] or by application of external gates [28]. With the application of gates, it is even possible to fill in single electrons in a very controlled way. Figure 5.3c shows a schematic picture of the band structure of an InAs SAQD sample, as used in [30]. There, an inverted modulation–doped AlGaAs–GaAs structure serves as a back contact. By application of a voltage between the back contact and a metallic front gate, the dots can be charged with single electrons, which then tunnel from the 2DES to the dots.

5.2.2 Electronic Ground State and Excitations

Macroscopic Picture

In this section, the ground state and the phenomenology of electronic elementary excitations in quantum dots will be discussed qualitatively. In the etched GaAs dots as well as in the InAs SAQD [31], the lateral *external* confining potential, which acts on a test electron inside the dot, is to a good approximation a two–dimensional harmonic–oscillator potential. In the deep–etched samples, the potential is determined by the homogeneously distributed ionized remote donors in the AlGaAs barrier, and negatively–charged surface states (see Fig. 5.4). By analytical calculations one can show that this lateral potential is in x and y direction (lateral directions) in good approximation parabolic [32]. We will therefore start our investigations by assuming that the corresponding single–particle energies of an electron in the conduction band

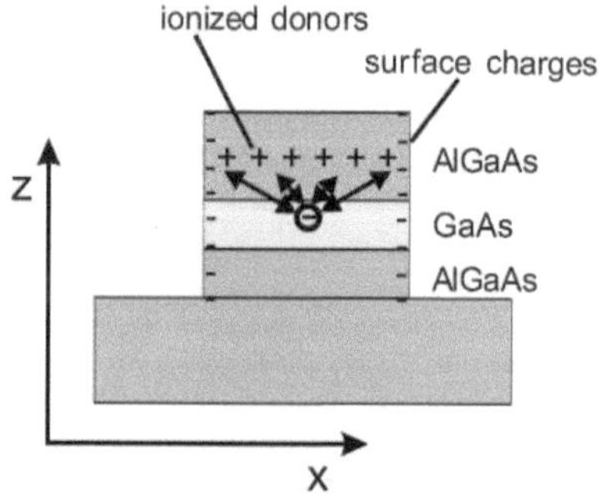

Fig. 5.4. Schematic cross section of a deep–etched GaAs–AlGaAs quantum dot. Negatively–charged surface states and ionized donors are indicated. Both charge distributions act on the test electron inside the GaAs layer

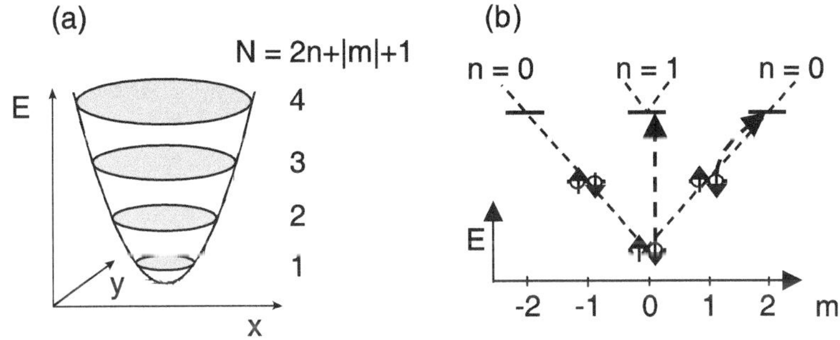

Fig. 5.5. (**a**) Schematic picture of the lateral potential of a quantum dot. (**b**) Energy levels of a quantum dot with parabolic potential. The *dashed arrows* indicate two possible excitations of single electrons

are given by the eigenvalues of a two–dimensional harmonic oscillator

$$E_{nm} = \hbar\Omega_0 \left(2n + |m| + 1\right) = N\hbar\Omega_0 \, . \tag{5.1}$$

Here, the two relevant quantum numbers are the radial quantum number $n = 0, 1, 2, ...$ and the azimuthal or angular–momentum quantum number $m = 0, \pm 1, \pm 2, \ldots$. We define a lateral quantum number $N = (2n + |m| + 1)$, which characterizes the $2N$-fold degenerate discrete levels, as sketched in Fig. 5.5a. Of course, the presence of many other electrons in the sample leads to a screening of the potential, which acts on the test electron. Theoretically, this is described, e.g., in a self–consistent Hartree calculation of the ground state (see Sect. 2.6). Thus, the effective lateral potential is flattened in the center of the dot, i.e., it gets more and more square–well–like. This results, in a mean field approach, in a decrease of the single–particle level spacings and to a lifting of degeneracies[1] For simplicity, we stay for the moment in the simple picture, where we assume that also the effective, or Hartree, potential is parabolic, which is a good approximation for the ground state of dots with very small electron numbers, only. In such a single–particle picture, the electronic excitations, which can be created in the inelastic light scattering process, are transitions of electrons from occupied to unoccupied levels, as, e.g., sketched in Fig. 5.5b. In Fig. 5.5b the energy levels for a parabolic quantum dot, which contains six electrons, is shown more detailed. The corresponding radial and angular–momentum quantum numbers are indicated. The vertical dashed arrow symbolizes an excitation, where the radial quantum number, n, is increased by 1, while the angular–momentum quantum number, m, is unchanged. The curved arrow, on the other hand, indicates a transition, where $\Delta m = 1$, and $\Delta n = 0$. Of course, in a real interacting system, the electronic excitations will be collective excitations, CDE's and

[1] This effect will be considered on a quantum–mechanical level in detail for quantum dots in Sect. 5.3.4.

SDE's, which are affected by Coulomb interaction. We will discuss the correct quantum–mechanical modelling of these collective excitations in quantum dots later, in Sects. 5.3.3, 5.3.4, and 5.4.3. Here, the aim is to introduce the important features in a qualitative model.

For large electron numbers, on the order of one hundred or more, it is instructive to leave the quasiatomistic picture described above for a while, and discuss the electronic excitations in terms of plasma oscillations of the electrons, to take account of the Coulomb interaction. For CDE's, in that case, some of the excitations are also called *confined plasmons*. For all following considerations, we assume that the external potential is parabolic. How can we imagine in such a system CDE's and SDE's? In Fig. 5.6 we make the attempt to visualize the lowest–energy CDE's and SDE's in a mesoscopic dot. The discs should indicate the lateral dot. Let us start with CDE's (plasmons). The lowest–energy CDE is a plasma oscillation where all electrons oscillate in phase in the quantum dot back and forth, independent of their spin directions. This situation is, for a fixed time, schematically shown in Fig. 5.6a, where a dark color should indicate a large induced electron density (due to an external time–dependent perturbation, as discussed in Sect. 3.3.1). White indicates an induced positive charge distribution. Hence, the induced electron density has in this case one node. Obviously, this excitation has a large dipole moment. It is also called the first confined plasmon or *Kohn's mode*, since for the case

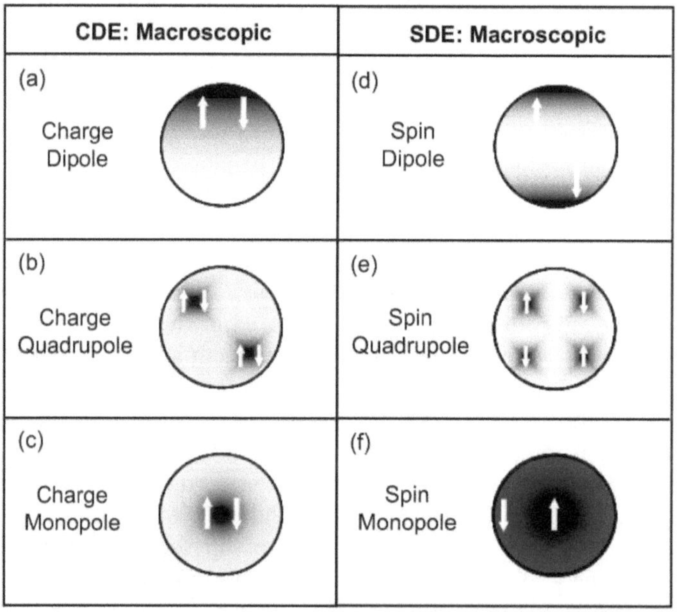

Fig. 5.6. Schematic drawings of the electron-density distributions for low–energy CDE's [(**a**)–(**c**)] and SDE's [(**d**)–(**f**)] in a disc–shaped two–dimensional quantum dot

of a parabolic external potential its energy equals exactly the quantization energy of the external potential and the energy of the lowest–energy dipole excitation is independent of the number of electrons, N, in the dot. This fact is known as the so called generalized Kohn's theorem [1]. The dipole excitation is characterized by a change of $\Delta M = 1$ of the *total* angular–momentum quantum number M, i.e., the angular momentum of the N–electron system. This will become more clear when we discuss the microscopic picture below. The CDE's with next higher energies are the quadrupole excitation corresponding to $\Delta M = 2$ (see Fig. 5.6b), and the monopole excitation with a change in total angular momentum of $\Delta M = 0$ (see Fig. 5.6c). This mode is also called a *breathing mode* due to the laterally symmetric oscillation of the induced charge density. The next higher excitation would again be a dipole excitation (not shown here), which is in addition accompanied by a change in the radial quantum number n. In all CDE's, the electrons with spin up and spin down oscillate in phase against the positively–charged background. This is schematically indicated in Fig. 5.6 by the white arrows. This means that the CDE's are strongly affected by the direct part of the Coulomb interaction, which results in a blueshift of the excitations. In contrast, in the corresponding SDE's, the electrons with spin up and spin down oscillate with a pase shift of π. For the spin–dipole excitation (see Fig. 5.7d), for example, this means that the center of mass of the charge does not oscillate. Thus, the energies of the SDE's are renormalized by exchange–correlation interaction, only, and are therefore redshifted. Figures 5.6e and 5.6f show the spin quadrupole and spin monopole SDE's, respectively. As introduced in Sect. 4.2.3, in experiments on Zincblende semiconductors, SDE's and CDE's can be distinguished by polarization selection rules [33]: CDE's are observed if the polarizations of the incoming and scattered light are parallel to each other (polarized geometry), and, for SDE's the polarizations have to be perpendicular (depolarized geometry).

Microscopic Picture

In the following we want to discuss how this macroscopic picture translates into the quasiatomistic one, which we introduced at the beginning. This makes sense for small electron numbers, only, otherwise it would be too complicated. The simplest interacting system one can think of, is a quantum dot with two electrons. In Figs. 5.7d, 5.7e, and 5.7f, the dipole, quadrupole, and monopole CDE's, respectively, are schematically shown for a two–electron quantum dot. More precisely, Figs. 5.7d–5.7f display pictorially the occupation of the Slater determinants, which predominantly contribute to the two–particle wavefunctions of the corresponding excited states. The exact many–particle wavefunction can always be written as an infinite series of single–particle Slater determinants. The ground state of a two–electron dot has a total angular momentum of $M = 0$, since both electrons occupy the s level with single–particle quantum number $m = 0$. From Fig. 5.7 one can

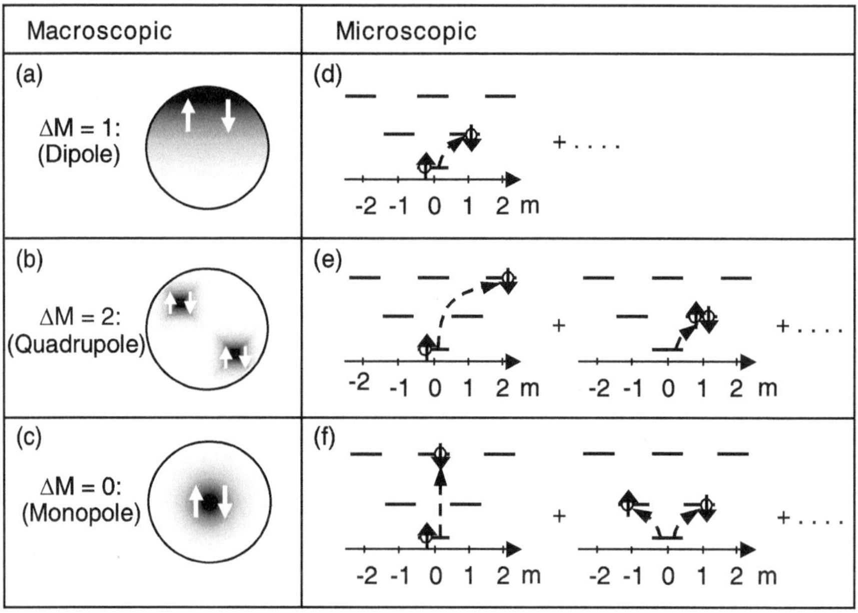

Fig. 5.7. Schematic drawings of the electron–density distributions for different confined plasmon modes [(**a**)–(**c**)]. Microscopic picture for the excitations of a parabolic quantum dot with two electrons [(**d**)–(**f**)]

see that for the dipole excitation the angular momentum of the excited state is $M = 1$, for the quadrupole excitation, $M = 2$, and, for the monopole excitation, $M = 0$. For the CDE's, as displayed in Fig. 5.7, the spin is preserved during the excitation ($S = S_z = 0$). For the excitation of SDE's, which are triplet excitations, $S = 1$, and S_z can take on values 1, -1, and 0. This means that spinflip transitions of electrons are involved in the SDE's. For zero magnetic field, $B = 0$, the three triplet excitations with $S_z = 0, \pm 1$, corresponding to the same excitation type (e.g., spin dipole, spin monopole, etc.), are energetically degenerate. As already mentioned, we expect the SDE's to have lower energies than the corresponding CDE's, since they feel no direct Coulomb interaction. As introduced in the first part in Sect. 4.2.3, the possibility to excite SDE's is a speciality of resonant inelastic light scattering. The observation of SDE's depends crucially on *resonant* scattering conditions.

5.3 GaAs–AlGaAs Deep-Etched Quantum Dots

In this chapter we are going to discuss inelastic light scattering experiments on modulation–doped GaAs–AlGaAs dots. In particular, in the first section we will investigate parity selection rules of collective excitations and the importance of resonant excitation for a correct theoretical description of the scattering process. In the second section, fine–structure effects in the spectrum of SDE's will be presented. The important role of resonant scattering conditions for the excitation of single–particle–like excitations will be emphasized in the third section by experimental examples and a theoretical modeling.

5.3.1 Parity Selection Rules in Quantum Dots

Figure 5.8 shows experimental inelastic light scattering spectra of quantum dots with approximately 200 electrons per dot [36]. In these experiments in backscattering geometry, the wave–vector transfer q, parallel to the plane of the dots, was close to zero. In each scattering configuration, depolarized and polarized, there is one mode visible. First of all, it is surprising that for an electron number as large as 200, the spectrum consists of only two active modes. On general grounds one can say that in a symmetric system the allowed modes have even parity since the inelastic light scattering process is a two–photon process (see Sect. 4.2.4). This is in contrast to direct absorption which is a one–photon process. Consequently, we can infer that the observed SDE and CDE are even–parity modes. Referring to the previous section, we conclude that the lowest–energy modes with even parity are the monopole modes with $\Delta M = 0$. The modes in Fig. 5.8 are labeled $(\Delta n, \Delta m)$ corresponding, respectively, to the changes in radial (n) and angular–momentum (m) quantum numbers of the involved single–particle transitions. The assignment of the modes is confirmed by experiments in a magnetic field, which will

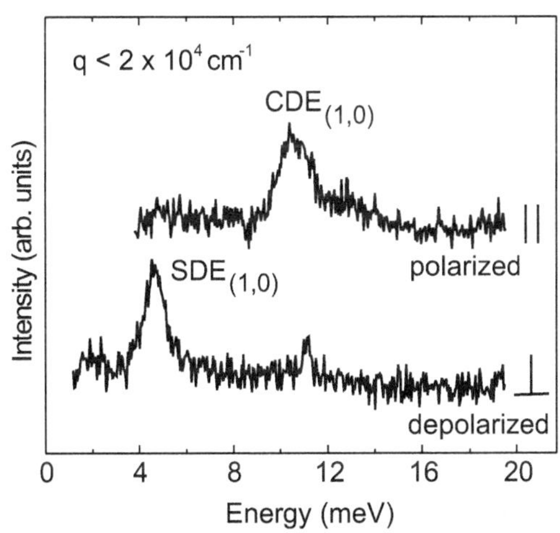

Fig. 5.8. Polarized and depolarized inelastic light scattering spectra of 240 nm GaAs–AlGaAs quantum dots with approximately 200 electrons per dot. The spectra were taken at $T = 12$ K [Reprinted from [36]. Copyright (2001), with permission from Elsevier]

be discussed in the following section. We want to note here that the polarized spectrum demonstrates the exclusion principle between FIR and Raman spectroscopy: the FIR–allowed mode $CDE_{(0,1)}$, the Kohn's mode, has odd parity and hence a large dipole moment (see previous section). It has for the sample displayed in Fig. 5.8 an energy of about 6 meV and is not visible at all in the Raman spectrum (see Fig. 5.8). On the other hand, the Raman–allowed monopole mode, $CDE_{(1,0)}$, has no dipole moment at all. Since the induced density of this mode has only nodes in radial direction [see Fig. 5.7c], it can also be regarded as a so called breathing mode, as already noted. So, the experiments demonstrate that for resonant inelastic light scattering on circularly–symmetric dots at approximately zero wave–vector transfer, the general parity selection rules hold.

In the following we want to elucidate these parity selection rules from a theoretical point of view in some more detail. Most of the calculations of inelastic light scattering spectra for low–dimensional electron systems have been carried out for nonresonant conditions, because in that case one does not have to deal with the complex valence–band states. Of course, this is a drastic simplification. Figure 5.9 shows a series of calculated resonant inelastic light scattering spectra for off–resonance and for different resonance conditions [36]. The RPA calculations were performed for a 30–electron quantum dot with $\hbar\Omega_0 = 8$ meV for the external parabolic potential[2]. In Fig. 5.9a, the laser energy was chosen to be far away from resonance, i.e., off–resonance conditions. Under experimental conditions, in this regime the scattered intensities are far too low to be observable. Note that the calculated spectrum in Fig. 5.9a is multiplied by a factor of 4000 in order to be of comparable strength with the resonant spectrum in Fig. 5.9c. One can see from Fig. 5.9a that in the off–resonance case the parity selection rules have completely changed: For curiosity, the FIR–allowed Kohn's mode, $CDE_{(0,1)}$, is the dominant mode! As the laser energy approaches resonance, the situation changes, and, under resonance conditions (Fig. 5.9c), the experimentally observed parity selection rules are confirmed, i.e., the monopole mode $CDE_{(1,0)}$ is the dominant mode in the polarized spectrum. This clearly demonstrates that for a correct description of the relative intensities of the excitations, the resonant scattering process is crucial.

An at least partial breakdown of the parity selection rules can be achieved by the transfer of a finite wave vector q parallel to the plane of the dot. Figure 5.10 shows a series of depolarized and polarized spectra for the same dots as in Fig. 5.8 for different wave–vector transfer q. As q increases, symmetry–forbidden modes [the spin dipole mode, $SDE_{(0,1)}$, and a higher dipole mode, $CDE_{(1,1)}$] gain relative intensity. Furthermore, the spin quadrupole mode, $SDE_{(0,2)}$, which for a parabolic effective potential would be energetically degenerate with the spin monopole mode, $SDE_{(1,0)}$, if interaction effects are neglected, becomes visible [12]. A so far unsolved puzzle is, why

[2] For more details about the calculation, see Sect. 5.3.3 and [26].

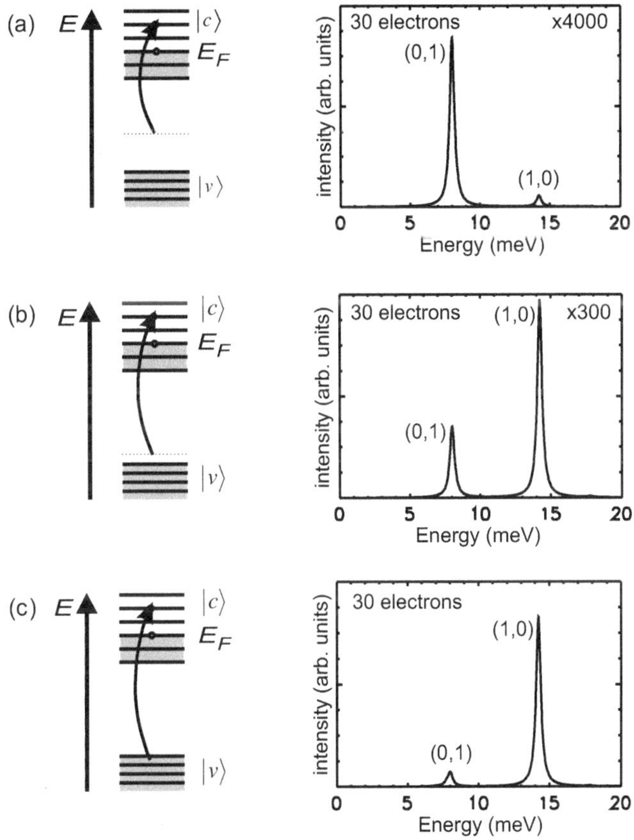

Fig. 5.9. Calculated inelastic light scattering spectra for a quantum dot with 30 electrons for (**a**) off–resonance, (**b**) close–to–resonance, and (**c**) resonant conditions. The schematic pictures show the different resonance conditions. The *curved arrows* indicate the laser energies

the violation of parity selection rules with wave–vector transfer q is much stronger for the SDE's than for the CDE's.

5.3.2 Fine Structure in Quantum Dots

In Chap. 5.2.2 we have introduced the collective excitations of quantum dots with large electron numbers in a qualitative macroscopic picture, and afterwards traced the resulting confined electron–density waves back to an atomistic picture, considering a 2–electron dot. However, even for large electron numbers, the mesoscopic artificial atom should in analogy to real 3D atoms be characterized by electronic shells and distinct quasiatomic orbitals. We have already used this assumption to label the excitations in Fig. 5.10. In order to justify this assignment and to highlight the internal structure, we have done

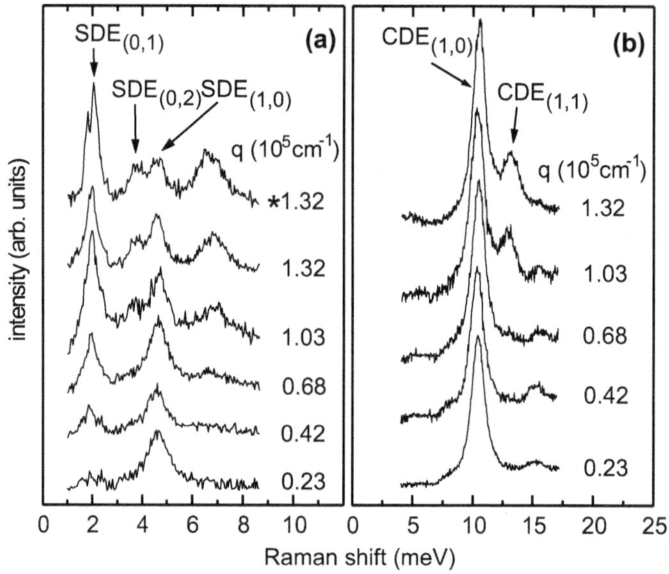

Fig. 5.10. (**a**) Depolarized, and (**b**) polarized inelastic light scattering spectra of GaAs quantum dots for different wave–vector transfer q, at a temperature of $T = 12$ K. The depolarized spectra in (a) were recorded at a laser energy $E_L = 1600.7$ meV, the spectrum marked with $(*)$ at $E_L = 1601.3$ meV [Reprinted with permission from [12]. Copyright (1998) the American Physical Society]

careful further analysis of the wave–vector and especially also magnetic–field dependence of the observed excitations [12, 14].

As discussed above, Fig. 5.10 shows depolarized (Fig. 5.10a) and polarized (Fig. 5.10b) spectra for different wave–vector transfer q, which were recorded at a laser energy where dominantly collective excitations are excited due to the TOP. Without proof, we have in the previous section labeled these excitations by the change in lateral quantum numbers of single–particle transitions, which dominantly contribute to the collective excitations. In this section we will now try to justify these assignments. Let us start by collecting some relevant experimental facts, which we can extract from Fig. 5.10. We see that the $\mathrm{SDE}_{(0,2)}$ occurs as an additional peak at the low–energy side of the $\mathrm{SDE}_{(1,0)}$ for large q. This additional peak can be resonantly enhanced by slightly changing the laser frequency (spectrum with an asterisk in Fig. 5.10a). We emphasize that the $\mathrm{SDE}_{(0,1)}$, $\mathrm{SDE}_{(1,0)}$ and $\mathrm{SDE}_{(0,2)}$ have essentially the same linewidths (≈ 0.5 meV) whereas the highest–energy peak in Fig. 5.10a has more than twice the linewidth, which means that it probably represents a superposition of two or more excitations, which are not individually resolved. Furthermore, we can see in Fig. 5.10a that the additional peak, $\mathrm{SDE}_{(0,2)}$, has almost exactly twice the energy of the $\mathrm{SDE}_{(0,1)}$.

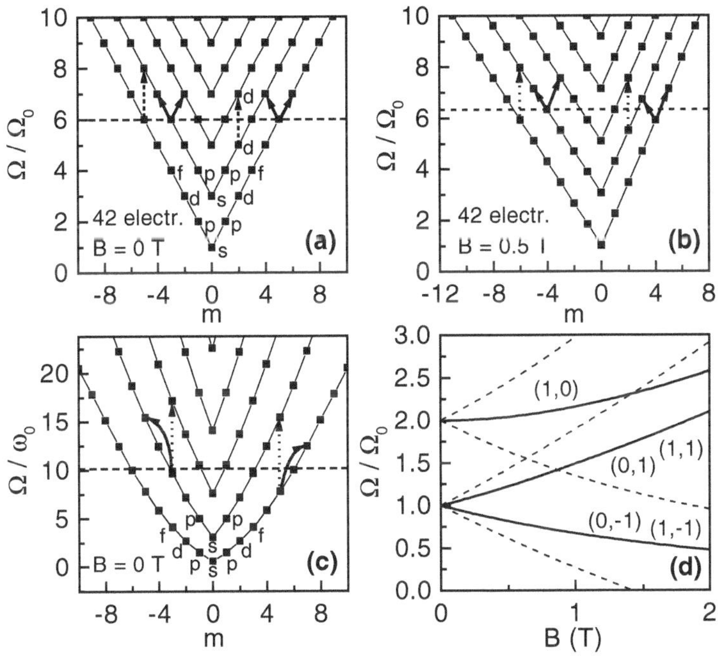

Fig. 5.11. Calculated energies for (**a**) a quantum dot with parabolic effective potential without and (**b**) with external magnetic field. (**c**) calculated energies in a dot with cylindrical symmetry and hard walls. (**d**) single–particle transition energies for a parabolic quantum dot with external magnetic field. The *solid lines* mark the experimentally observed transitions, the *dashed lines* give the forbidden next higher transitions. For the two *dashed lines* starting at $\Omega/\Omega_0 = 1$ the corresponding transitions are $(-1, \pm 3)$ and $(2, \pm 3)$, for the *dashed lines* starting at $\Omega/\Omega_0 = 2$, $(0, \pm 2)$ and $(2, \pm 2)$. In a hard–wall potential, the $(0, \pm 2)$ and $(2, \pm 2)$ are shifted to lower energies, which results in a fine structure. Ω_0 is the quantization energy of the parabolic potential. $\omega_0 = \hbar/(2m^*)(\pi/a)^2$ characterizes the hard–wall potential, where a is the dot diameter in that case [Reprinted with permission from [12]. Copyright (1998) by the American Physical Society]

In order to learn something about the internal structure of these excitations in the quasiatomic system, we analyze our observations in a simple model: As we have learned from the investigations in Sect. 3.3, exchange–correlation effects are expected to be small in our structures so that we can neglect for a moment many–particle interactions and treat the SDE's as simple single–particle transitions. As a first approach, we also start with a two–dimensional parabolic potential for the effective lateral potential of our quantum dot. The eigenenergies of this two–dimensional harmonic oscillator were given in (5.1). The corresponding energy spectrum is displayed in Fig. 5.11a. Each filled square in Fig. 5.11 represents an allowed energy state, which can be occupied by two electrons due to spin degeneracy. Here we have

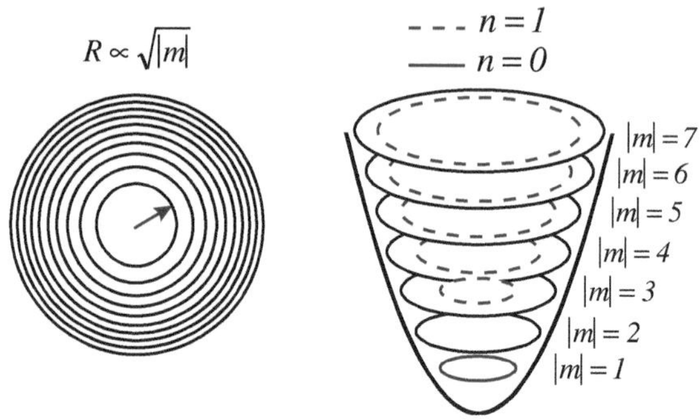

Fig. 5.12. Sketch of the classical electron orbits in a quantum dot with parabolic lateral potential

considered, for simplicity, 42 electrons per dot, which makes no qualitative difference in the interpretation. The solid lines connect points which belong to the same radial quantum number n, where $n = 0, 1, 2, \ldots$ rises from the bottom to the top. All points which can be connected by a horizontal line in Fig. 5.11a belong to the same lateral quantum number $N = (2n + |m| + 1)$ and represent the $2N$–fold degenerate discrete levels of the dot. Some states are marked with $'s'$, $'p'$, etc., to indicate the character of the quasiatomic orbitals as, e.g., quasi–s or quasi–p orbitals. Actually, these orbitals have not the same symmetry as orbitals of real atoms, therefore we call them *quasi*. In a classical picture, the electrons with finite angular–momentum quantum number m move on circular orbits within the dot, where the radius R is proportional to the square root of $|m|$ as shown in the left part of Fig. 5.12. With increasing radial quantum number n, the set of orbits increases in energy as can be seen in the right part of Fig. 5.12. The circular movement of the electrons represents a current which produces a magnetic moment $\boldsymbol{M}$, perpendicular to the plane in which the current flows. This magnetic moment can have different directions depending on the sign of the quantum number m. If we now apply a magnetic field, $\boldsymbol{B}$, perpendicular to the x–y–plane, $\boldsymbol{B}$ can be oriented either parallel or antiparallel to $\boldsymbol{M}$. This has as a consequence that states with $\boldsymbol{M}$ parallel to $\boldsymbol{B}$ decrease in energy, while states with $\boldsymbol{M}$ antiparallel to $\boldsymbol{B}$ increase in energy. This is shown in Fig. 5.11b, where a finite magnetic field has been considered. In general, the energy spectrum in a parabolic quantum dot in dependence on a perpendicular magnetic field is given by [37]

$$E_{mn}(B) = (2n + |m| + 1)\,\hbar\sqrt{\Omega_0^2 + \left(\frac{\omega_c}{2}\right)^2} + m\hbar\frac{\omega_c}{2}\,, \tag{5.2}$$

where $\omega_c = eB/m^*$ is the cyclotron resonance frequency. Within the model described so far, the lowest–energy SDE – $\mathrm{SDE}_{(0,1)}$ – consists of transitions from the highest occupied to the first unoccupied level of the dot, i.e.,

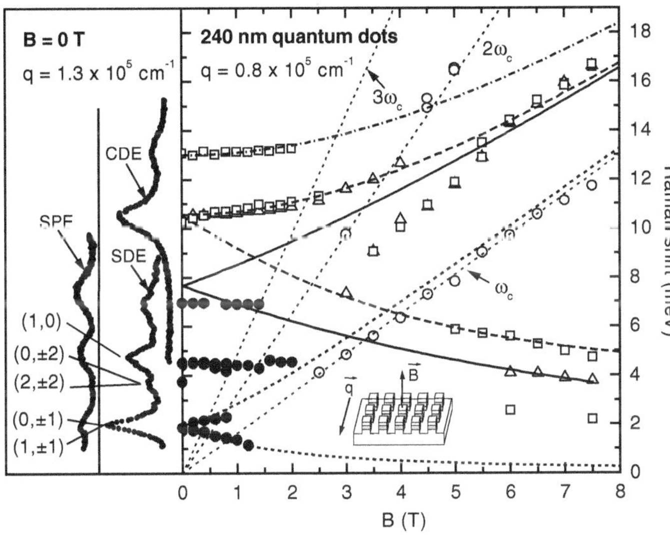

Fig. 5.13. Magnetic field dispersions of the observed modes in a quantum dot sample. In the *left* panel spectra of SDE's and CDE's for $B = 0$ are displayed. The spectrum of SPE's was recorded at a laser energy $E_L = 1561$ meV under conditions of extreme resonance. The spectra of SDE's and CDE's were taken at $E_L = 1587$ meV. The numbers in the brackets, $(\Delta n, \Delta m)$, give the changes in radial (Δn) and azimuthal (Δm) quantum numbers for the observed transitions [Reprinted with permission from [12]. Copyright (1998) by the American Physical Society]

$\Delta N = 1$. In principle, several of such transitions are possible with different changes in radial and azimuthal quantum numbers. In Fig. 5.11a only some transitions with $(\Delta n = 0, \Delta m = \pm 1)$ and $(\Delta n = 1, \Delta m = \pm 1)$ are indicated by solid arrows. We claim that only these transitions predominantly contribute to the observed $\mathrm{SDE}_{(0,1)}$. This can be deduced from magnetic–field dependent measurements: Fig. 5.13 displays the magnetic–field dependence of the SDE's (solid symbols) and the CDE's (open symbols). In the left panel of Fig. 5.13, characteristic spectra of SDE's and CDE's, at $B = 0$ and large wave–vector transfer q, are displayed for illustration. For comparison, we have also included a spectrum, which is dominated by SPE's under conditions of extreme resonance (see following Sect. 5.3.3), where in this spectrum we have substracted a hot–luminescence background. We find that in a magnetic field the usual polarization selection rules for plasmons are weakened. Excitations which are most prominent in polarized geometry are marked by squares, and excitations which are dominant in depolarized configuration are marked by triangles. We want to concentrate here on the SDE's because the behavior of the CDE's is well known from FIR transmission experiments [38]. In Fig. 5.13 we can see that the $\mathrm{SDE}_{(0,1)}$ mode splits with finite magnetic field. If we compare this splitting with the expected splitting of the transitions

($\Delta n = 0, \Delta m = \pm 1$) and ($\Delta n = 1, \Delta m = \pm 1$), calculated with (5.2) (see Fig. 5.11d), this gives the correct result. Also the dotted lines starting at $E = 2$ meV and $B = 0$ in Fig. 5.13 are calculated with (5.2) for the above given ($\Delta n, \Delta m$). Transitions with $\Delta m \geq 2$ would result in larger splittings (see, e.g., dashed lines in Fig. 5.11d), which are experimentally not observed. Therefore, this confirms the spin–monopole character of this lowest–energy mode. Within this parabolic model, the splitting between the two branches with $\Delta n, |\Delta m| \leq 1$ is exactly ω_c. The slight deviation of the upper branch from the theoretical curve can be due to a nonparabolicity of the effective potential or to the remaining collective character of the SDE's. The collective nature of the SDE's furthermore manifests itself in the vanishing of the excitations at relatively small B. This is an effect of Landau damping: At finite B, all degeneracies are lifted and a wealth of single–particle lines cross the energies of the SDE's so that they get Landau damped. For illustration, the forbidden next higher transitions are displayed in Fig. 5.11d (dashed lines).

Also for the $\text{SDE}_{(1,0)}$, in principle several energetically degenerate transitions with different ($\Delta n, \Delta m$) are possible as can be seen in Fig. 5.11a. The experiments show neither a splitting nor a shift with magnetic field (Fig. 5.13). Among all possible transitions, only transitions with ($\Delta n = 1, \Delta m = 0$), as indicated by dotted arrows in Fig. 5.11a, show such characteristics (see also Fig. 5.11d). Transitions with $|\Delta m| \geq 2$ would exhibit splittings $\geq 2\omega_c$ (see also dotted lines in Fig. 5.11d). This verifies that this mode is a spin monopole mode. In these experiments, the wave–vector transfer was restricted to $q \approx 0.8 \times 10^5$ cm^{-1} [12]. At this relatively small q, we can not follow all peaks, especially the SPE's and the additional peak, $\text{SDE}_{(0,2)}$, in the spectrum of SDE's, in a magnetic field. Nevertheless, from the energetic position of the additional peak, which lies below the position of the $\text{SDE}_{(1,0)}$, we can deduce that it consists of transitions with $\Delta n = 0, 2$ and $\Delta m = \pm 2$, although these transitions are in a parabolic potential energetically degenerate with the transitions with $\Delta n = 1$ and $\Delta m = 0$ (dashed arrows in Fig. 5.11a): From geometrical considerations we can estimate about 200 electrons per dot in our structures. At this relatively large number of electrons screening is important and the real effective potential would certainly deviate from a parabolic shape (cf. Sect. 5.3.4). As an extreme limit we have calculated the eigenenergies for a square–well potential with cylindrical symmetry. The real dot potential should be in between these limits. The corresponding energy spectrum is shown in Fig. 5.11c. Indeed, from Fig. 5.11c we can see that transitions with $|\Delta m| = 2$ (solid arrows) have lower energies than transitions with ($\Delta n = 1, \Delta m = 0$) (dotted arrows). Furthermore, transitions with $|\Delta m| = 2$ have almost exactly twice the energy of transitions with $|\Delta m| = 1$, if we move along lines with constant n in Fig. 5.11d. This is observed in the experiment as can be seen in Fig. 5.10a, and verifies that the additional mode below the spin monopole mode is a spin quadrupole excitation.

Concluding this section, investigations with applied magnetic field and varying wave–vector transfer q allowed us to identify transitions between different electronic shells of the quasiatomic systems, which predominantly contribute to the observed low–energy SDE's. We find that these excitations are formed by transitions with $\Delta n, |\Delta m| \leq 1$ (spin–dipole and spin–monopole excitations), whereas transitions with $|\Delta m| = 2$ (spin–quadrupole excitation) can only be observed at relatively large q and form a fine structure [12, 14].

5.3.3 The Important Role of Extreme Resonance

Experimental Evidence for Single–Particle Excitations

The collective excitations described so far obey the well–known polarization selection rules, i.e., CDE's appear in polarized, and SDE appear in depolarized spectra. In our experiments on low–dimensional electron systems [11] we found that under conditions of extreme resonance, when the laser energy is close to the fundamental band gap, excitations appear which show stronger resonance enhancements than the collective excitations and which appear in both polarization configurations (cf. Sect. 4.2.1 for experiments on quantum wells). In quantum wells, such excitations, which were interpreted as SPE's, were first observed by Pinczuk et al. [39]. In the next section we will try to highlight the character of such excitations, which seem to have very small energy renormalizations due to Coulomb interaction, in quasiatomic systems from a theoretical point of view in more detail. Here, we want to elucidate first the experimentally observed behavior.

In Fig. 6.5, experimental spectra are shown for different laser energies E_L. For E_L well above the fundamental bandgap (E_L = 1587 meV), the collective SDE's and CDE's can be observed. As mentioned before, these spectra obey the well–known polarization selection rules. As the laser energy is tuned towards the fundamental bandgap, excitations labeled SPE's appear, which are visible in both scattering geometries. From the energy shifts between the excitations, the relative strengths of the many-particle interactions can be derived [11]. In the following we want to compare this experimental finding qualitatively to resonant TDLDA calculations.

Theoretical Modeling

For the calculations [26] we assume an external parabolic potential, as written in (5.1) on page 92. A very important point is that due to the circular symmetry of the dot, contributions to excitations with different Δm completely separate [26] and an excitation is determined by a distinct Δm. Vice versa, excitations with different Δm can be calculated separately. Since there is more than one electron inside a dot, the external potential is partially screened and hence the *effective* potential is *not* parabolic. We start our calculations by determining the ground state of the many–electron dot by solving the

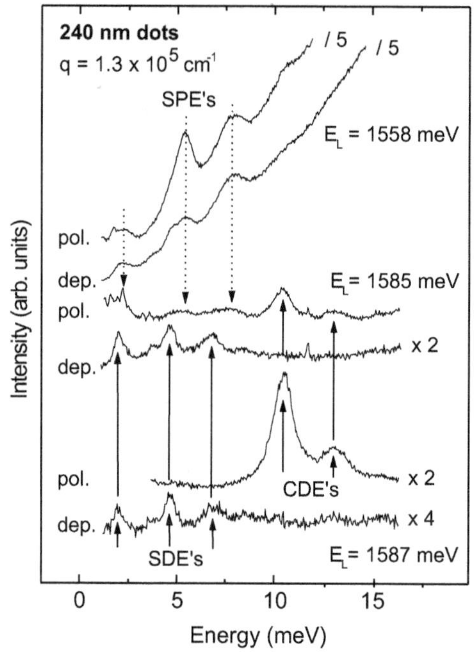

Fig. 5.14. Experimental polarized and depolarized Raman spectra for different laser energies E_L in a quantum-dot sample with about 200 electrons per dot. The temperature was $T = 12$ K

Kohn–Sham Hamiltonian (see Sects. 2.4.2 and 2.6) self–consistently. The radial part reads

$$H = -\frac{\hbar^2}{2m^*}\left(\frac{1}{r}\frac{d}{dr} + \frac{d}{dr^2} - \frac{m^2}{r^2} + V_{\text{ext}}(r) + V_{\text{H}}(r) + V_{\text{XC}}(r)\right) , \qquad (5.3)$$

where $m^* = 0.07$ m$_0$ is the effective mass of GaAs, $V_{\text{ext}}(r)$ is the external parabolic potential, $V_{\text{H}}(r)$ is the self–consistent Hartree potential, and $V_{\text{XC}}(r)$ is the LDA exchange–correlation potential. Details of the calculations can be found elsewhere [26]. The nonparabolic effective potential, $V_{\text{ext}}(r) + V_{\text{H}}(r) + V_{\text{XC}}(r)$, results in a lifting of degeneracies of the energy eigenvalues compared to the 2D harmonic oscillator. As an example, the allowed energy eigenvalues for a dot with 12 electrons, i.e., three completely filled quasiatomic shells, and $\hbar\Omega_0 = 6$ meV for the external potential are shown in Figs. 5.15b and c. Due to spin degeneracy, two electrons can occupy an allowed energy state. In Fig. 5.15b, transitions between single–particle levels with $(\Delta n, \Delta m) = (0, 1)$ are indicated, and, in Fig. 5.15c, transitions with $(1, 0)$ are shown. From the previous section we remember that the transitions $(0, 1)$ form the collective dipole modes, and the transitions $(1, 0)$ the monopole modes. Due to

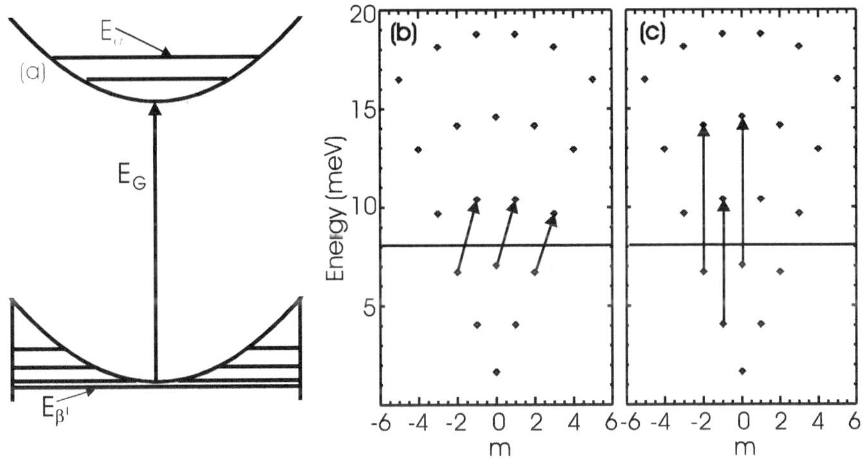

Fig. 5.15. (**a**) Schematic drawing of the conduction–band and valence band potentials and energy levels in a quantum dot. (**b**), (**c**): Self–consistently calculated energy states in a quantum dot with 12 electrons. (b) shows FIR–allowed $\Delta m = 1$ transitions (Kohn mode), (c) Raman–allowed $\Delta m = 0$ transitions

the mentioned lifting of degeneracies, all displayed single–particle transition energies are *nondegenerate*.

As demonstrated before, the experimental experience shows that in most cases electronic excitations in low–dimensional semiconductor structures can be observed only under specific interband resonance conditions. This means that the valence–band states play a crucial role in the scattering process. We have modeled resonant Raman spectra by including the valence band [26]. To do so, we have calculated, as a first step, the quantized states of the heavy–hole band. Heavy–hole–light–hole mixing is neglected, for simplicity. The energy states and self–consistent potentials are shown schematically in Fig. 5.15a. According to [23], the Raman cross section is given by the imaginary part of a generalized correlation function (see also Sect. 4.2.2)

$$\frac{d^2\sigma}{d\Omega d\omega} \propto -\Im \frac{i}{\hbar} \int_{-\infty}^{\infty} dt\theta(t) \exp(i\omega t) \langle [N^\dagger(t), N(0)] \rangle \ . \tag{5.4}$$

$\hbar\omega$ is the energy of the elementary excitation, and N is the generalized pair operator (cf. (4.25) on page 69)

$$N = \sum_{\alpha,\beta} \gamma_{\alpha\beta} c_\beta^\dagger c_\alpha \ . \tag{5.5}$$

$c_\beta^\dagger$ and c_α are, respectively, creation and annihilation operators of single–particle states in the conduction band. When the laser energy is close to the resonance, the scattering amplitudes are given by (cf. (4.26) on page 69)

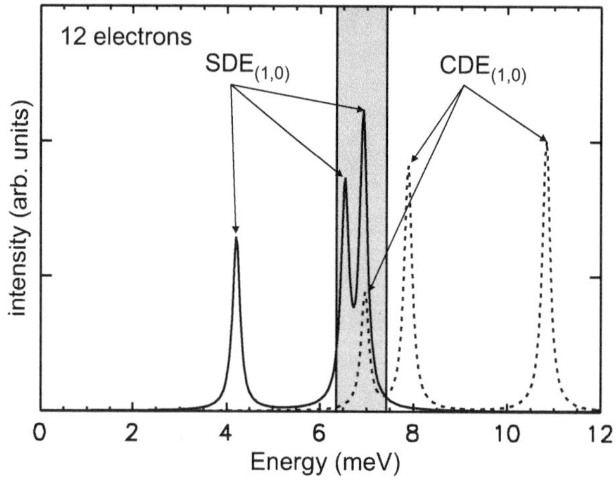

Fig. 5.16. Calculated resonant inelastic light scattering spectra for a 12–electron quantum dot. The *solid* and *dashed lines* are the depolarized and polarized spectra, respectively. The *gray–shaded* area marks the region of $(1,0)$ single–particle transitions

$$\gamma_{\alpha\beta} = \frac{1}{m_0} \sum_{\beta'} \frac{\langle\alpha|\, \boldsymbol{e}_I\, \boldsymbol{p}\, e^{i\boldsymbol{k}_I \boldsymbol{r}}\, |\beta'\rangle\langle\beta'|\, \boldsymbol{p}\, \boldsymbol{e}_S\, e^{i\boldsymbol{k}_S \boldsymbol{r}}\, |\beta\rangle}{E_\alpha - E_{\beta'} - \hbar\omega_I - i\eta}\,, \tag{5.6}$$

where E_α and $E_{\beta'}$ are single–particle energies of states in the conduction band and the valence band, respectively. $\boldsymbol{e}_I$ ($\boldsymbol{e}_S$) and $\boldsymbol{k}_I$ ($\boldsymbol{k}_S$) are the polarization vectors and wave vectors of the incoming (scattered) light, $\boldsymbol{p}$ is the momentum operator, and $|\beta'\rangle$ denotes a state in the valence band. With this formalism we have calculated resonant Raman spectra for quantum dots with up to 30 electrons [26]. The electron numbers in the experimental structures, which were discussed in the previous section, are on the order of 100, so we certainly can not quantitatively compare the results.

In Fig. 5.16, calculated spectra for a 12–electron quantum dot with $\hbar\Omega_0 = 6$ meV, and $E_I - E_G = 12$ meV, i.e., a laser energy very close to the fundamental bandgap E_G, are shown. We want to note again that these are just model calculations, since we have considered only one heavy–hole band and neglected heavy–light–hole mixing. In Fig. 5.16 we have calculated the Raman–allowed $(1,0)$ modes. There are three CDE and three SDE collective modes for these strong–resonance conditions. All excitations consist of *coupled* $(1,0)$ transitions. As can be seen from Fig. 5.15c, for a 12–electron dot with three completely filled shells, there are three energetically different $(1,0)$ transitions (the vertical arrows drawn in Fig. 5.15c). This results in three charge–density and three spin–density modes for polarized and depolarized configurations, respectively. Surprisingly, two modes of each type, charge density and spin density, are inside or very close to the energetic region of the single–particle transitions between Kohn–Sham states (gray–shaded region in

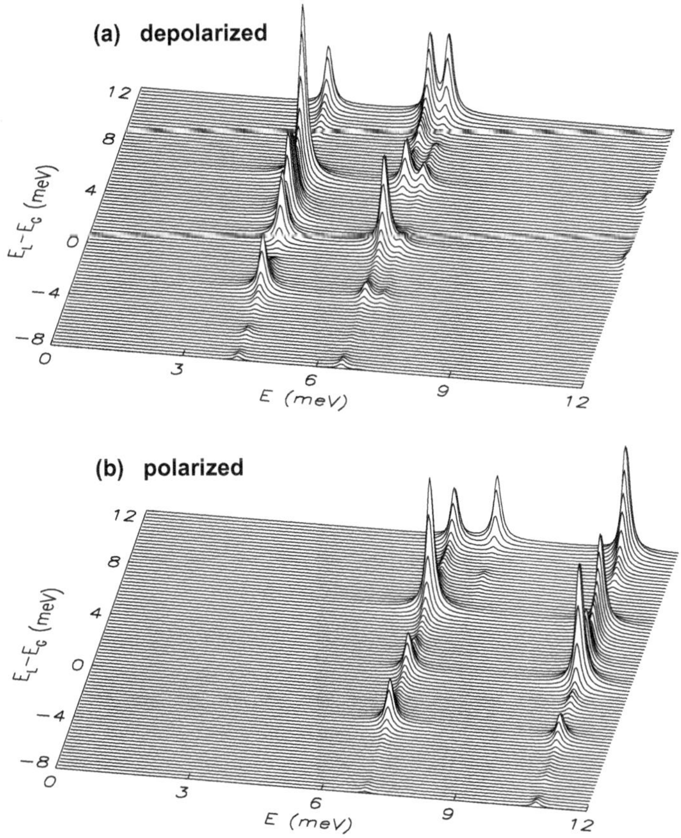

Fig. 5.17. Calculated resonant Raman spectra for a 12–electron quantum dot for different detunings, $E_L - E_G$, between laser energy and fundamental bandgap

Fig. 5.16). In a macroscopic picture, these modes are "out–of–phase" modes of the oscillations of the induced densities. This means that depolarization–field effects are small for these modes and hence the collective shifts are small. On the other hand, the strongly shifted principal modes are "in–phase" modes with strong depolarization–field effect.

In Fig. 5.17, series of resonant depolarized (Fig. 5.17a) and polarized (Fig. 5.17b) spectra are displayed for different detunings, $E_L - E_G$, between the energy of incoming photons and the fundamental bandgap E_G. Maxima due to resonance with different confined hole levels can be seen.

We will see in the next section that for a 6–electron quantum dot, i.e., two completely filled quasiatomic shells, two of such single–particle–like excitations exist. We assume that for many–electron dots, as studied by us experimentally, the broad SPE's evolve from these "out–of–phase" modes which are, strictly speaking, also collective charge–density and spin–density modes.

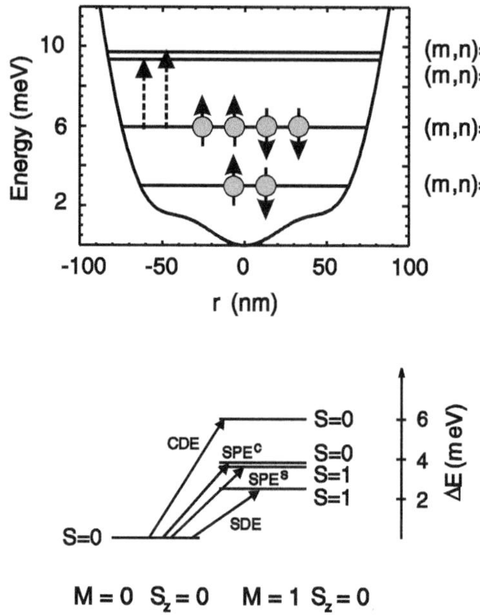

Fig. 5.18. Comparison of electronic levels in a quantum dot with 6 electrons (symbolized by *gray–shaded circles*), as derived in a Kohn–Sham calculation (*upper panel*) and an exact numerical diagonalization (*lower panel*). m and n are the single–particle angular–momentum and radial quantum numbers, respectively. M and S denote, respectively, the total angular momentum and total spin

This might also be the reason that the experimentally observed excitations are visible in both polarization configurations.

5.3.4 Calculations for Few-Electron Quantum Dots

We will complete in this section the investigation of single–particle–like excitations in quantum dots by presenting an intriguing comparison [40] of Kohn–Sham calculations with results of an exact many–body treatment by numerical diagonalization. With the latter technique, the electron–electron interaction is treated exactly. Furthermore, we give a microscopic picture for the electronic excitations by a superposition of Hartree–Fock states. Figure 5.18 shows theoretical results for a quantum dot with 6 electrons, i.e., two filled quasiatomic shells and $\hbar\Omega_0 = 6$ meV. The upper panel of Fig. 5.18 shows the resulting self–consistent potential and the first three electronic levels derived from a Kohn–Sham calculation. The arrows indicate the two low–energy SPE's which are possible in this single–particle picture. To derive the collective excitations starting from this basis, one has to perform, e.g., a TDLDA calculation as discussed before. On the lower panel of Fig. 5.18, the lowest many–particle states, resulting from an exact numerical diagonal-

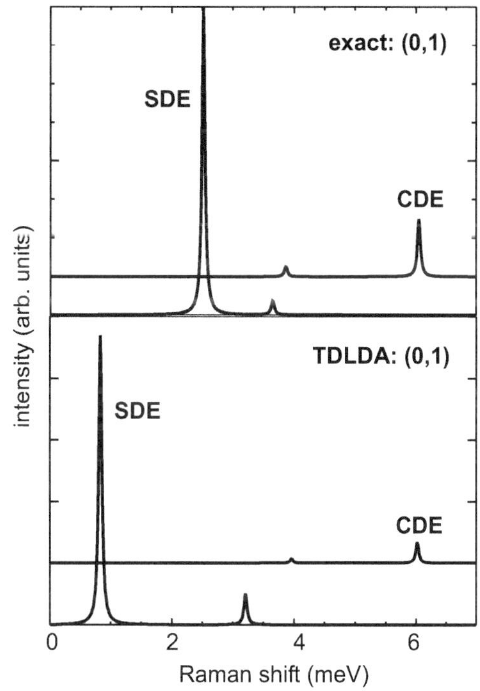

Fig. 5.19. Comparison of exact calculations with TDLDA calculations of a 6–electron quantum dot

ization, are shown. Here, the many–body interaction is treated exactly and therefore this calculation directly yields the correct excitation energies of the interacting six–electron system. One can identify the collective SDE's and CDE's as singlet–triplet ($\Delta S = 1$) and singlet–singlet ($\Delta S = 0$) transitions, respectively. Very similar to the results presented in the previous section, there are two excitations, which are very close together in energy. One with spin–density (labeled $\mathrm{SPE^S}$) and one with charge-density character (labeled $\mathrm{SPE^C}$). These transitions have almost exactly the same energies as the spacing between the Kohn–Sham single–particle levels, shown in the upper panel (see vertical arrows in Fig. 5.18). From this we can draw the conclusion that even for only six electrons per quantum dot, electronic excitations with energies close to unrenormalized single–particle transitions can be expected. As shown above, for 12 electrons per quantum dot, we found four corresponding transitions with energies close to single–particle transitions.

Figure 5.19 shows a comparison of spectra calculated by exact diagonalization (upper panel) with TDLDA calculations (lower panel). This unambiguously shows that the single–particle–like transitions obtained above are not an artefact of the TDLDA, since they appear in both theories. One conclusion from the calculations displayed in Fig. 5.19 is that in the TDLDA

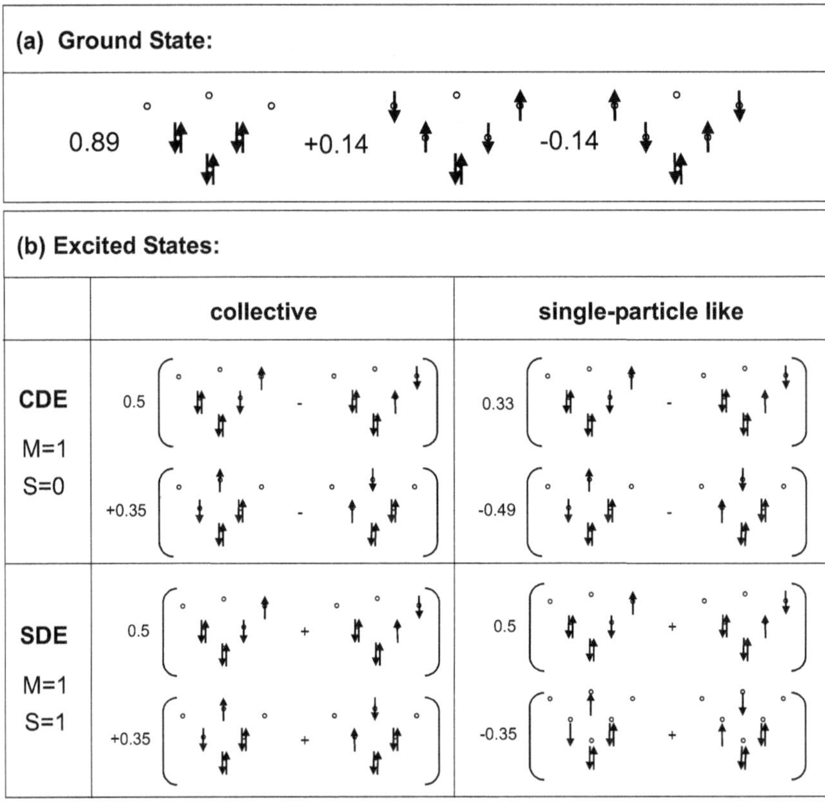

Fig. 5.20. Schematic presentation of the Hartree–Fock states which dominantly contribute to the many–electron ground state (**a**) and excited states (**b**). Up and down arrows symbolize electrons with spin up or spin down, respectively

correlation effects are underestimated since the SDE's are at lower energies than in the exact treatment.

In the following we want to give a more microscopic picture, again by expanding the exact many–body states in a Hartree–Fock basis. The exact states can be described by an infinite series of Hartree–Fock states, i.e., Slater determinants. In Fig. 5.20, the dominant Hartree–Fock determinants for the ground state (Fig. 5.20a) and the excited states (Fig. 5.20b) are shown schematically, by displaying the occupancy of single–particle states in each determinant. The triplet states ($S = 1$) are three–fold degenerate ($S_z = 1, 0, -1$) in energy for $B = 0$. Therefore, we have considered only $S_z = 0$ in our calculation. As a consequence, in Fig. 5.20b no spinflip transitions are present. For the triplet states with $S_z = 1, -1$, in each determinant a spinflip transition would occur.

Concluding this section about experiments and theory on GaAs–AlGaAs quantum dots with rather large electron numbers, the main results shall be emphasized. These are: (i) in mesoscopic dots, the parity selection rules for inelastic light scattering on symmetric systems are confirmed, and for a correct theoretical treatment of these selection rules, the inclusion of the valence band is crucial. (ii) Under conditions of extreme resonance, excitations are observed, whose energies are close to unrenormalized single–particle transitions. (iii) Resonant TDLDA calculations show that these excitations are indeed collective CDE and SDE, where collective effects partly cancel due to out–of–phase oscillations of the induced charge and spin densities.

5.4 InAs Self-Assembled Quantum Dots

5.4.1 Few–Electron Quantum–Dot Atoms

The fascination of InAs SAQD comes from the fact that these objects have very small size – laterally a few tens of nanometers and a few nanometers in hight. This causes relatively large quantization energies of several 10 meV, which make even room temperature devices feasible. Furthermore, it has been demonstrated that by the application of metallic gates, InAs SAQD can be charged with single electrons [28]. Hence, they can serve as few–electron quantum–dot atoms. In contrast to the mesoscopic dots, which were discussed in the previous section, in the SAQD the quantization energy is larger than the typical Coulomb interaction energies. Therefore, the quantization should dominate, and the Coulomb interaction is expected to cause corrections to this.

In this section, some of the first inelastic light scattering experiments on InAs SAQD, which contain small numbers of electrons, will be discussed [30, 41]. For the experiments which will be described below, a special sample design, where a Q2D electron system was used as a back contact, was applied (see Fig. 5.3 on page 90). The charging of the dots can be monitored in situ by capacitance measurements. The capacitance trace, displayed in Fig. 5.21, shows that by varying V_{Gate} the s and the p shells of the quantum dots can be

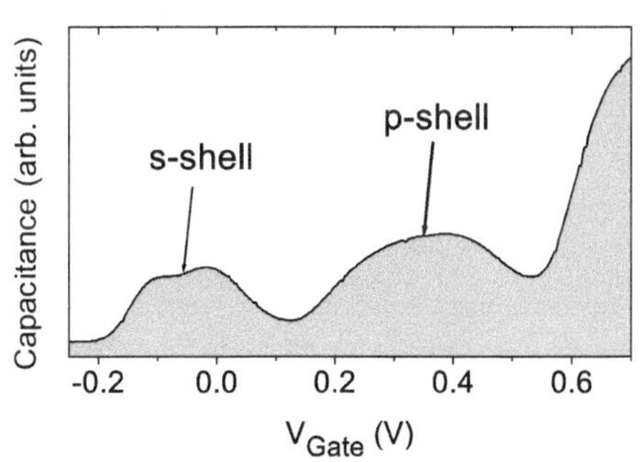

Fig. 5.21. Differential capacitance of InAs SAQD versus applied gate voltage

charged with electrons. The doublet structure around $V_{Gate} = -0.05$ V arises from the charging of the s shell with the first and the second electron, and, at the broad plateau at positive V_{Gate}, the p shell is loaded with 4 electrons. The strong increase of the signal for $V_{Gate} > 0.6$ V is due to the charging of the wetting layer. At this point it is not clear whether or not a bound d shell, very close to the wetting layer continuum, exists in the dots.

5.4.2 Electronic Excitations in InAs SAQD

For the inelastic light scattering experiments, resonant excitation was achieved by tuning the laser to the $E_0 + \Delta$ gap of the InAs SAQD ($\sim$ 1.65 eV), which is far above the fundamental PL transition energies of the structure ($\sim 1.1 - 1.2$ eV). Figure 5.22 displays polarized inelastic light scattering spectra for two selected situations: For the upper spectrum, the dots are charged, on the average, with two electrons, and, for the lower spectrum the s and the p shells are completely occupied. The spectra show two dominant bands, labeled as A and B. The two sharp lines at $E = 33.4$ meV and $E = 36.6$ meV are due to the TO- and LO-phonon excitations of the GaAs bulk material in the structure. In a simplified single–particle picture, the band A in Fig. 5.22 was interpreted [30] as a CDE due to transitions of electrons from the s to the p shell (s–p transitions) of the quasiatoms, and the band B as p–d transitions: It was assumed that 2 to 3 confined single–particle energy levels exist in the quantum dots, which is schematically shown for $N = 2$ and

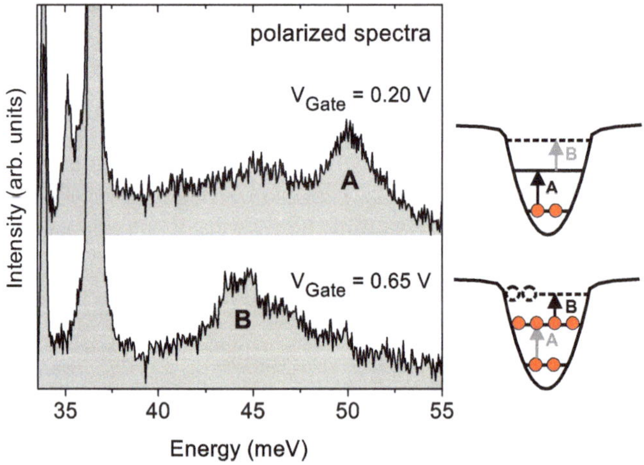

Fig. 5.22. Polarized inelastic light scattering spectra for two different gate voltages V_{Gate}, at a temperature of $T = 5$ K. The *right* panel sketches the charging state of the dots. Because it is not clear whether a bound d shell exists, the state has been drawn with *dashed lines*. The *black* (*gray*) *vertical arrows* indicate possible allowed (forbidden) transitions of electrons [Reprinted with permission from [30]. Copyright (2003) by the American Physical Society]

$N = 6$ to 8 electrons per quantum dot in the drawings on the right–hand side of Fig. 5.22. The black vertical arrows indicate possible transitions of single electrons. The self–assembled quantum dots stick out of an InAs wetting layer. Therefore, the dot potentials are flattened at the edges due to the wetting–layer continuum. This leads to the situation that the single–particle transition, sketched as B in the inset of Fig. 5.22, has a smaller energy than the transition A. At that point it can not definitely be excluded that the transition B is an excitation of electrons from the p shell into the wetting layer. This would be the case if there would be no bound d shell. It is obvious from Fig. 5.22 that the band A dominates, if the p shell is empty, and the band B is most pronounced if the p shell is completely filled. In this situation, the band A is strongly suppressed because of the Pauli exclusion principle.

5.4.3 Comparison with Exact Calculations

To analyze the dependence of the observed excitations on the electron number N in the dots in more detail, Fig. 5.23a displays a series of experimental spectra for $N = 2 \ldots 6$ electrons in the dots. Interestingly, with increasing number of electrons in the p shell, in Fig. 5.23a, (i), a shift of A to lower energies, and, (ii), a broadening of the transition can be observed. This is somehow intriguing, since naively one would expect for a CDE a shift to higher energies if the number of electrons, i.e., the electron density, in the

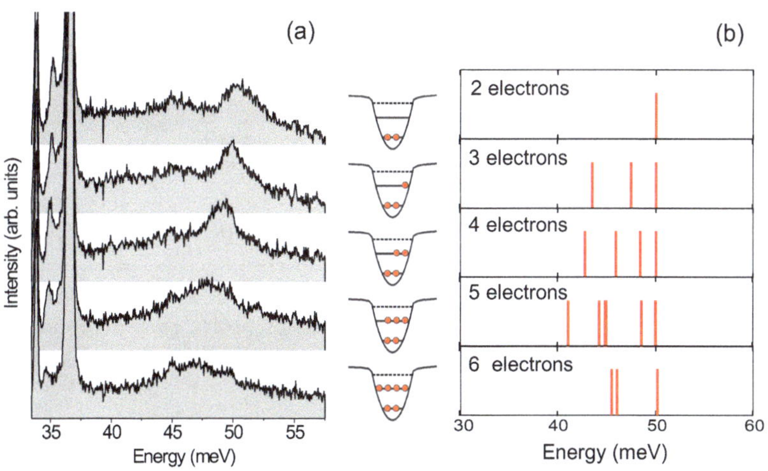

Fig. 5.23. (**a**) Polarized inelastic light scattering spectra for gate voltages $V_{Gate} =$ 0.2 V, 0.25 V, 0.35 V, 0.45 V, 0.55 V (from *top* to *bottom*). The inset shows the charging states of the dots. (**b**) Calculated energies of low–energy collective excitations of a two–dimensional parabolic quantum dot for different electron numbers in the dot

quantum system is increased, if one would argue in terms of a depolarization shift. Obviously, in order to explain the experimentally–observed effect, we have to study the Coulomb interaction in the quantum dot in more detail. For this end we consider the following model Hamiltonian

$$H = \sum_{i=1}^{N} \left[\frac{\boldsymbol{p}_i^2}{2m^*} + \frac{m^*}{2} \Omega_0^2 \mathbf{r}_i^2 \right] + \frac{e^2}{4\pi\epsilon\epsilon_0} \sum_{i \neq j}^{N} \frac{1}{|\mathbf{r}_i - \mathbf{r}_j|} \tag{5.7}$$

of a two–dimensional quantum dot with a parabolic confining potential in lateral direction $[\mathbf{r} = (r_x, r_y)]$ with quantization energy $\hbar\Omega_0 = 50$ meV. This potential certainly deviates from the more realistic shape as sketched in Fig. 5.22. Thus, the model will not be able to include the effect of the lower energy of band B, as compared to that of band A. However, our aim is to get qualitative results, which show the correct tendencies of the dependence of energies on electron number. For the effective mass m^*, the InAs bulk value $m^* = 0.024\ m_0$, and for the dielectric constant $\epsilon = 15.15$ were used. Qualitatively, the results do, of course, not depend on the particular choice of these parameters. By exact numerical diagonalization we have determined the low–energy excitations of the N-electron quantum dot, where during the excitation the total spin is preserved [30, 41]. This selects the excitations which we expect in polarized inelastic light scattering spectra (CDE's). Figure 5.23b exhibits the calculated excitation energies for electron numbers $N = 2 \ldots 6$ in the quantum dot. First, one can see that, independent of N, there is always an excitation at the energy $\hbar\Omega_0 = 50$ meV of the external confining potential. This is a direct consequence of the so called generalized Kohn's theorem [1], which says that in a parabolic potential, the energy of the center–of–mass oscillation of all electrons – i.e., the charge–dipole excitation – is independent of the number of electrons. For $N > 2$, however, in our calculations additional mode energies appear below the energy of the Kohn's mode. In the previous section we saw that for the case of 6 electrons, these additional excitations are close to single–particle transition energies. We will come back to this interesting point below. The most important result of Fig. 5.23b, concerning our experiments, is that with increasing N, the spectral weight of the low–energy excitations shifts to lower energies. This might explain the shift and broadening of the band A as observed in the experiments (Fig. 5.23a) to be due to additional excitations at lower energies which can not be individually resolved in the ensemble experiment. Furthermore, as discussed in Sect. 5.2.2, the Kohn's mode should from the point of view of parity selection rules not be observable in inelastic light scattering experiments, since it is a dipole excitation with odd parity. Therefore, it is possible that the excitations observed in the experimental spectra in Fig. 5.23 are solely due to the additional excitations at lower energies.

In the remainder we want to highlight the microscopic origin of the additional modes at lower energies in more detail. In order to make the situation not too complicated, we discuss the low–energy excitations for 3 electrons

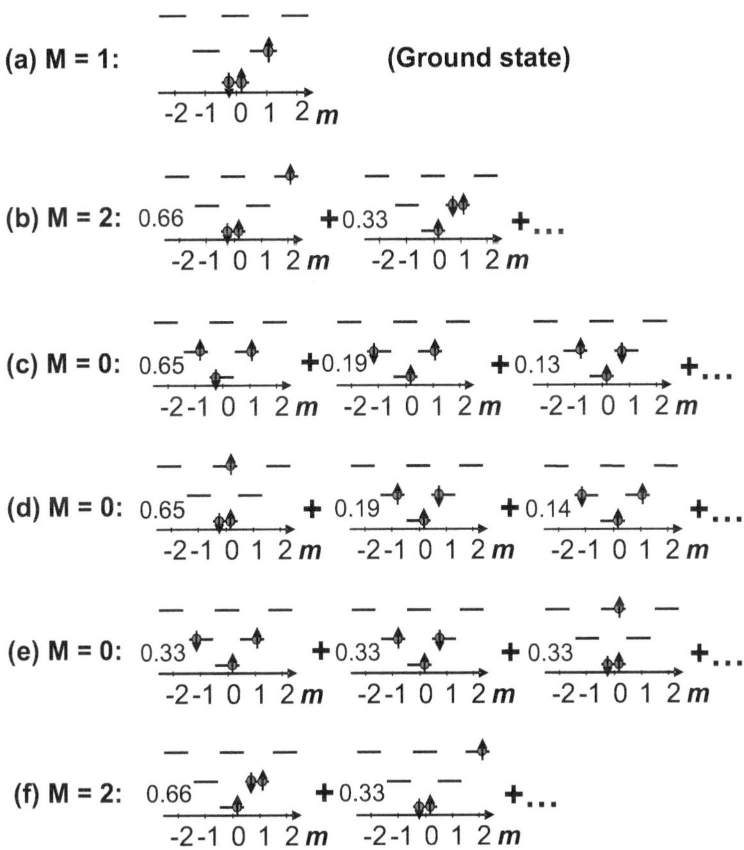

Fig. 5.24. Schematic picture of occupation in the most important Slater determinants, which contribute to the exact many–body wave function of (**a**) the ground state with $M = 1$ and $S = S_z = \frac{1}{2}$, and, (**b**)–(**e**) its low–energy excited states A similar picture with the same eigenenergies could be drawn for $M = -1$, and/or $S_z = -\frac{1}{2}$ for the ground state. The numbers give the occupation probabilities of the respective determinants. M is the total angular–momentum quantum number, and m is the single–particle angular momentum [Reprinted with permission from [30]. Copyright (2003) by the American Physical Society]

inside a dot. In Fig. 5.23b we can see that for 3 electrons there are two additional modes below the Kohn mode, with energies of 43.5 meV and 47.4 meV. The exact many–particle wavefunction can be expanded in a series of Slater determinants. Figure 5.24 displays pictorially the occupation of the dominant Slater determinants, which contribute to the ground state (Fig. 5.24a) and the lowest excited states (Figs. 5.24b–e) of a quantum dot which contains 3 interacting electrons. For 3 electrons, the ground state has already a finite total angular momentum of $M = 1$ (Fig. 5.24a), which separates into

$M = M_{CM} + M_{rel}$, with $M_{CM} = 0$ for the center–of–mass motion, and $M_{rel} = 1$ for the relative motion of the electrons. For the ground state, the complete angular momentum is in the relative motion of the electrons. This is because in the ground state the electrons try to avoid each others as much as possible in order to minimize the Coulomb energy. We find that the Kohn mode is twofold degenerate (Figs. 5.24b and 5.24c), with either $M = 2$, or $M = 0$, corresponding to a change of $\Delta M_{CM} = \pm 1$ in the center–of–mass motion, while $M_{rel} = 1$ stays constant. This verifies that, macroscopically, as discussed in Sect. 5.2.2, the Kohn mode is an in–phase oscillation (center–of–mass oscillation) of all electrons in the dot. Microscopically, the dominant contributions here are excitations where either the electron in the p shell is excited into the d shell (Fig. 5.24b), or a superposition of three equivalent configurations (Fig. 5.24c). On the other hand, a change of angular momentum only in the *relative* motion, $\Delta M_{rel} = \pm 1$, leads to a lowering of the excitation energies (total $M = 0$, Fig. 5.24d, and, $M = 2$, Fig. 5.24e). Microscopically, an excitation where the two electrons in the p shell have a paired spin leads to the strongest energy reduction (Fig. 5.24e). In any case, the difference in total angular–momentum, ΔM, between excited states and the ground state is $\Delta M = \pm 1$. This also concludes our considerations of single–particle–like excitations of the previous sections: excitations, which appear at lower energies than the Kohn's mode are due to relative motion of electrons.

A qualitative understanding of the excitation–energy reduction with increasing N can also be gained by first order perturbation theory in the Coulomb operator. Such calculations qualitatively reproduce our exact results. The perturbation approach shows that the Coulomb interaction is strongest if both electrons occupy the s state and decreases if one or both electrons occupy a p orbital. This is due to the larger spatial extension of the p orbitals compared to the s orbital [42]. An excitation is obtained, e.g., by transferring an s electron to a p orbital, which leads to a reduction of the Coulomb interaction between this electron and all others. Clearly this reduction will increase with increasing number of electrons. It shall be also noted that the important qualitative result of this model, i.e., the appearance of additional collective modes at lower energies seem to be quite general for quantum dots: Similar calculations for a three–dimensional spherical quantum dot with a step–like potential yield also a reduction of the lowest–energy excitations with increasing N.

So, one of the key findings in the presented experiments [30, 41] is that, obviously, in inelastic light scattering experiments on few–electron quantum dots one is able to observe more complex excitations than only the center–of–mass oscillation of all electrons, which is usually observed in far–infrared absorption [43, 44, 45, 46]. This is quite general for inelastic light scattering, and the reasons are mainly (i) the two–photon nature of the process, and, (ii) the possibility to overcome parity selection rules by a finite wave–vector transfer into the system [14]. Even though, in the present experiments, the additional low–energy excitations are not individually resolved but lead to a broadening of the observed band.

References

1. P. Maksym and T. Chakraborty: Phys. Rev. Lett. **65**, 108 (1990)
2. M. A. Reed, J. N. Randall, R. J. Aggarwal, R. J. Matyi, T. M. Moore, and A. E. Wetsel: Phys. Rev. Lett. **60**, 535 (1988)
3. W. Hansen, T. P. Smith III, K. Y. Lee, J. A. Brum, C. M. Knoedler, J. M. Hong, and D. P. Kern: Phys. Rev. Lett. **62**, 2168 (1989)
4. C. Sikorski and U. Merkt: Phys. Rev. Lett. **62**, 2164 (1989)
5. T. Demel, D. Heitmann, P. Grambow, and K. Ploog: Phys. Rev. Lett. **64**, 788 (1990)
6. A. Lorke and J. P. Kotthaus: Phys. Rev. Lett. **64**, 2559 (1990)
7. B. Meurer, D. Heitmann, and K. Ploog: Phys. Rev. Lett. **68**, 1371 (1992)
8. K. Bollweg, T. Kurth, D. Heitmann, V. Gudmundsson, E. Vasiliadou, P. Grambow, and K. Eberl: Phys. Rev. Lett. **76**, 2774 (1996)
9. R. Strenz, U. Bockelmann, F. Hirler, G. Abstreiter, G. Böhm, and G. Weimann: Phys. Rev. Lett. **73**, 3022 (1994)
10. D. J. Lockwood, P. Hawrylak, P. D. Wang, C. M. Sotomayor Torres, A. Pinczuk, and B. S. Dennis: Phys. Rev. Lett. **77**, 354 (1996)
11. C. Schüller, G. Biese, K. Keller, C. Steinebach, D. Heitmann, P. Grambow, and K. Eberl: Phys. Rev. B **54**, R17304 (1996)
12. C. Schüller, K. Keller, G. Biese, E. Ulrichs, L. Rolf, C. Steinebach, D. Heitmann, and K. Eberl: Phys. Rev. Lett. **80**, 2673 (1998)
13. G. Biese, C. Schüller, K. Keller, C. Steinebach, D. Heitmann, P. Grambow, and K. Eberl: Phys. Rev. B **53**, 9565 (1996)
14. Christian Schüller, in Festkörperprobleme / Advances in Solid State Physics, Ed. B. Kramer (Vieweg Braunschweig Wiesbaden 1999) Vol. 38, p. 167
15. S. Das Sarma and Daw-Wei Wang: Phys. Rev. Lett. **83**, 816 (1999)
16. G. Danan, A. Pinczuk, J. P. Valladares, L. N. Pfeiffer, K. W. West, and C. W. Tu: Phys. Rev. B **39**, 5512 (1989)
17. A. O. Govorov: J. Phys. Condens. Matter **9**, 4681 (1997)
18. O. Steffens and M. Suhrke: Phys. Rev. Lett. **82**, 3891 (1999)
19. M. Pi, M. Barranco, A. Emperador, E. Lipparini, Ll. Serra: Phys. Rev. B **57**, 14783 (1998)
20. Ll. Serra, M. Barranco, A. Emperador, M. Pi, E. Lipparini: Phys. Rev. B **59**, 15290 (1999)
21. E. Lipparini, M. Barranco, A. Emperador, M. Pi, Ll. Serra: Phys. Rev. B **60**, 8734 (1999)
22. A. Mooradian: Phys. Rev. Lett. **20**, 1102 (1968)
23. F. A. Blum: Phys. Rev. B **1**, 1125 (1970)
24. M. Sassetti and B. Kramer: Phys. Rev. Lett. **80**, 1485 (1998)
25. D. W. Wang, A.J. Millis, S. Das Sarma: Phys. Rev. Lett. **85**, 4570 (2000)
26. C. Steinebach, C. Schüller, and D. Heitmann: Phys. Rev. B **59**, 10240 (1999)
27. For a recent review see: Daniel Gammon and Duncan G. Steel: Physics Today 2002, October, p. 36, and references therein

28. see, e.g., H. Drexler, D. Leonard, W. Hansen, J. P. Kotthaus, and P. M. Petroff: Phys. Rev. Lett. **73**, 2252 (1994); R. J. Warburton et al.: Nature **405**, 926 (2000); F. Findeis et al.: Phys. Rev. B **63**, 121309(R) (2001)
29. L. Chu, A. Zrenner, M. Bichler, G. Böhm, and G. Abstreiter: Appl. Phys. Lett. **77**, 3944 (2000)
30. T. Brocke, M.-T. Bootsmann, M. Tews, B. Wunsch, D. Pfannkuche, Ch. Heyn, W. Hansen, D. Heitmann, and C. Schüller: Phys. Rev. Lett. **91**, 257401 (2003)
31. A. Wojs, P. Hawrylak, S. Fafard, L. Jacak: Phys. Rev. B **54**, 5604 (1996)
32. U. Merkt, Physica B **189**, 165 (1993)
33. D. Hamilton, A. L. McWhorter, Light Scattering Spectra of Solids, Ed. G. B. Wright (Springer, New York, 1969) p. 309
34. E. Burstein, A. Pinczuk, D. L. Mills: Surf. Sci. **98**, 451 (1980)
35. A. Pinczuk, L. Brillson, and E. Burstein: Phys. Rev. Lett. **27**, 317 (1971)
36. C. Schüller, C. Steinebach, and D. Heitmann: Solid State Commun. **119**, 323 (2001)
37. V. Fock: Z. Phys. **47**, 446 (1928)
38. see D. Heitmann and J.-P. Kotthaus in *Physics Today*, June 1993, p. 56, and references therein
39. A. Pinczuk, S. Schmitt–Rink, G. Danan, J. P. Valladares, L. N. Pfeiffer, and K. W. West: Phys. Rev. Lett. **63**, 1633 (1989)
40. Christoph Steinebach, Christian Schüller, and Detelf Heitmann: Phys. Rev. B **61**, 15600 (2000)
41. T. Brocke, M.-T. Bootsmann, M. Tews, B. Wunsch, D. Pfannkuche, Ch. Heyn, W. Hansen, D. Heitmann, and C. Schüller: Phys. Rev. Lett. **91**, 257401 (2003)
42. R. J. Warburton, B. T. Miller, C. S. Dürr, C. Bödefeld, K. Karrai, J. P. Kotthaus, G. Medeiros-Ribeiro, P. M. Petroff, and S. Huant: Phys. Rev. B **58**, 16221 (1998)
43. M. Fricke et al., Europhys. Lett. **36**, 197 (1996)
44. A. Wojs and P. Hawrylak, Phys. Rev. B **53**, 10841 (1996)
45. S. Sauvage, P. Boucaud, R. P. S. M. Lobo, F. Bras, G. Fishman, R. Prazeres, F. Glotin, J. M. Ortega, and J.-M. Gérard: Phys. Rev. Lett. **88**, 177402 (2002)
46. S. Hameau, J. N. Isaia, Y. Guldner, E. Deleporte, O. Verzelen, R. Ferreira, G. Bastard, J. Zeman, and J. M. Gérard: Phys. Rev. B **65**, 085316 (2002)

6 Quantum Wires: Interacting Quantum Liquids

6.1 Introduction

In 1989, the first inelastic light scattering experiments on electronic excitations in quantum wires were reported [1, 2]. Since then, a number of experimental papers appeared about, e.g., many–particle interactions and selection rules in those systems [3, 4, 5, 6, 7, 8, 9] and investigations with applied external magnetic field [10, 11, 12]. All these experiments were performed on lithographically–defined GaAs–AlGaAs structures. Consequently, the lateral sizes of these structures were on the order of 100 nm, or at least not much below [8, 9]. Unlike for the case of quantum dots, there is no well–established method of self–organized growth of modulation–doped quantum wires. During the past few years, Carbon nanotubes have evolved as new and alternative quantum–wire structures. So far, the main focus in the investigation of those very promising quantum structures by optical experiments has been on phonon excitations [13]. Phonon Raman spectroscopy has greatly helped in unveiling the topological structure of Carbon nanotubes [13]. An interesting further method to produce very narrow wires with atomic–layer precision is the so called cleaved–etched overgrowth (CEO) [14]. However, with CEO it is difficult to grow very large arrays of wires, which would be necessary to get enough signal strength in inelastic light scattering experiments. Hence, there are so far no reports of inelastic light scattering experiments on CEO wires, though these might be promising structures for high–sensitivity experiments. As mentioned, most of the existing experimental reports are on lithographically–defined GaAs–AlGaAs quantum wires with rather mesoscopic widths. Hence, in those experimental structures, typically several Q1D subbands are occupied with electrons. In this chapter we will discuss both, experiments and calculations on such samples. The main focus will be on the microscopic origin of confined plasmons and interesting internal interaction effects in a magnetic field. These experimental results are described well within the RPA, i.e., a Fermi–liquid theory, as we will see later.

Much *theoretical* research on Q1D electron systems has been triggered by the fact that in the so called Tomonaga–Luttinger model [15] one is able to treat the Coulomb interaction in a *strictly* one–dimensional system exactly. This model relies on the approximations that the one–dimensional subband dispersion is linearized in the vicinity of the Fermi wave vector, and backscat-

Christian Schüller: *Inelastic Light Scattering of Semiconductor Nanostructures*
STMP **219**, 121–143 (2006)
DOI 10.1007/3-540-36526-5_6

tering events are neglected. For this case, the so called bosonization technique allows one to treat the Coulomb interaction exactly. An important result of this theory is that in a Luttinger liquid, all elementary electronic excitations are *intrinsically* collective [16], i.e., electrons are no defined quasiparticles. This is a very important difference to the Fermi–liquid theory. Hence, in a Luttinger liquid, there are no SPE's, and one should expect collective SDE's and CDE's as the only electronic excitations. There has, however, been some debate, whether or not a Luttinger liquid can be realized in a real quantum wire sample [17, 18]. One would assume that a Q1D quantum wire has to be at least in the quantum limit, i.e., only one subband is occupied by electrons, in order to be close to a Luttinger liquid. There is only one report in literature – of F. Perez et al. [8, 9] – where the authors report about the realization of very narrow GaAs–AlGaAs quantum wires, which are supposed to be in the quantum limit. However, experimentally it seems to be difficult to decide if either the Fermi liquid, or the Luttinger liquid theory is better suited to describe the experimental observations [9]. The dilemma here is that the Q1D plasmon wave–vector dispersion, which can be measured by inelastic light scattering, is in both theories exactly the same [17]. In Sect. 6.4 we will discuss experiments on very narrow, lithographically–defined structures, which are close to the quantum limit. However, the experimental proof of a Luttinger liquid still remains a challenge.

6.2 Electronic Elementary Excitations in Quantum Wires

6.2.1 Ground State and Excitations

Similar to quantum dots, in deep–etched quantum wires, the lateral confining potential is determined by the homogeneously distributed ionized remote donors in the AlGaAs barrier, and negatively–charged surface states, as sketched in Fig. 6.1. By analytical calculations one can show that this lateral potential, which acts on a test electron within the structure (*bare* potential), is for the case of dots [19] in x and y direction, and, in wires, for the direction perpendicular to the wires (in the following y direction) in good approximation parabolic. Of course, the presence of other electrons in the one–dimensional channel screens the interaction of the test electron with the remotely localized charges, and the effective potential will deviate from a parabolic shape. A positive effect of the negative surface charges, however, is that due to Coulomb repulsion the *electronic* width, a, of the Q1D electron system is smaller than the geometrical width w (see Fig. 6.1). The length $(w-a)/2$ is called depletion length. Its value depends on the structure of the sample and may also depend on experimental conditions, as the illumination strength. For optical experiments, the depletion length is typically a few tens of nanometers [8].

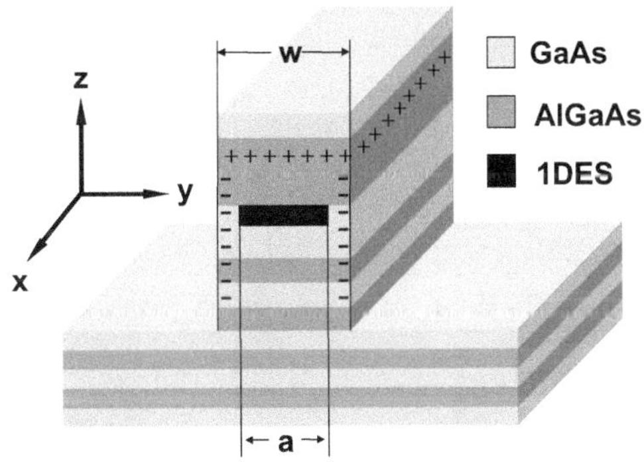

Fig. 6.1. Schematic cross section of a deep–etched quantum wire perpendicular to the wire direction

Without loss of generality, we assume in the following for qualitative considerations that the electron motion in the wires is in y direction quantized by a parabolic potential and in x direction quasifree. The two quantum numbers, which then characterize this problem, are a discrete 1D subband quantum number $i = 0, 1, 2, \ldots$ for the quantized movement, and a quasicontinuous quantum number k_x. The corresponding energy eigenvalues are given by

$$E_{i,k_x} = \hbar\Omega_0 \left(i + \frac{1}{2} \right) + \frac{\hbar^2 k_x^2}{2m^*} \; . \tag{6.1}$$

Ω_0 is the quantization energy and m^* the effective mass. This situation is sketched in Fig. 6.2a for the case of 2 occupied Q1D subbands. In a single–particle picture, the electronic excitations, which can be created in the inelastic light scattering process, are transitions of electrons from occupied to unoccupied states. Such SPE's are sketched in Fig. 6.2a for different wave–vector transfers, q_x, in wire direction. The striking difference of Q1D systems, as compared to Q2D electrons (cf. Sect. 3.1), is that the dispersion relations of the Q1D subbands are one–dimensional parabolas in k_x direction. We remember that for Q2D systems the dispersion relations are paraboloids in k_x and k_y. Therefore, the corresponding single–particle continua for Q1D electron systems differ from those of Q2D systems (cf. Fig. 3.2 on page 43). In Fig. 6.2b, the lowest single–particle continua (grey–shaded regions) for a Q1D electron system with two occupied subbands are schematically shown. Since there are two occupied Q1D subbands, there are two intrasubband continua ($\Delta i = 0$). The difference to the case of a Q2D system is that for finite q_x the continua start at finite energy. This is a consequence of the one–dimensional subband dispersion (cf. Sect. 3.1). There is also a forbidden region in the continuum of the $\Delta i = 1$ intersubband continuum, which is due to the blocking of transitions from $i = 0$ to $i = 1$, since there are occupied states in the $i = 1$ subband. For n occupied Q1D subbands, we would have $2n - 3$ such regions

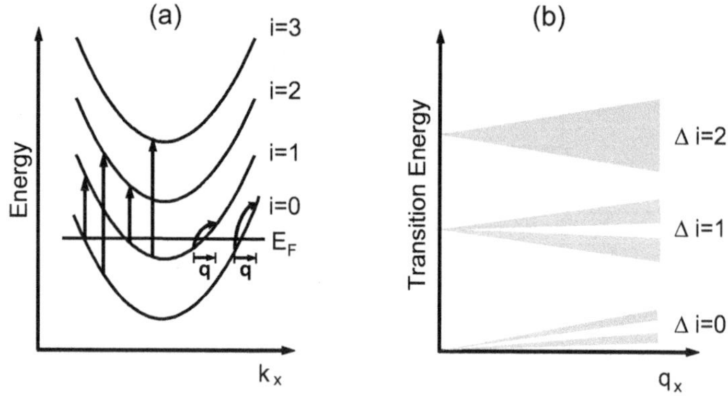

Fig. 6.2. (**a**) Q1D subband dispersion of one-dimensional subbands in a quantum wire. (**b**) Single-particle continua

in the $\Delta i = 1$ intersubband continuum, if $n > 1$. In experiments on quantum wires with two occupied subbands, this forbidden region was detected in the spectrum of $\Delta i = 1$ intersubband transitions by A. R. Goñi et al. [5].

In Sect. 5.2.2 we have introduced the collective elementary excitations of quantum dots in a simple macroscopic model, where we discussed the oscillations of the induced densities. This can also be done for the case of quantum wires: In a Q2D electron system, the energy of the charge–density wave, which propagates freely parallel to the layer plane is proportional to the square root of the wave vector $q_{||}$ (intraband plasmon, cf. Sect. 3.8). This is sketched in Fig. 6.3a. In a semiclassical picture, one can assume that due to the lateral confinement in y direction, the inplane Q2D charge wave is quantized. The characteristic of the first confined mode, which we will in the following call CDE_1, is one node in the induced density (see Fig. 6.4). This would be the Kohn's mode of the wire, i.e., the dipole–active mode (for a parabolic bare potential). Consequently, with increasing index of the confined mode, the number of nodes in the induced density increases, too (cf. Fig. 6.4). However, in the wire, the charge wave can still propagate freely along the x direction. Therefore, the confined plasmon modes exhibit a wave–vector dispersion, if a finite wave vector is transferred parallel to the wire direction. This wave–vector dependence is sketched in Fig. 6.3b, and was calculated, e.g., in a semiclassical hydrodynamical model by Eliasson et al. [20]. The microscopic quantum–mechanical origin of the localized CDE's in quantum wires will be discussed in detail below in Sect. 6.3.1.

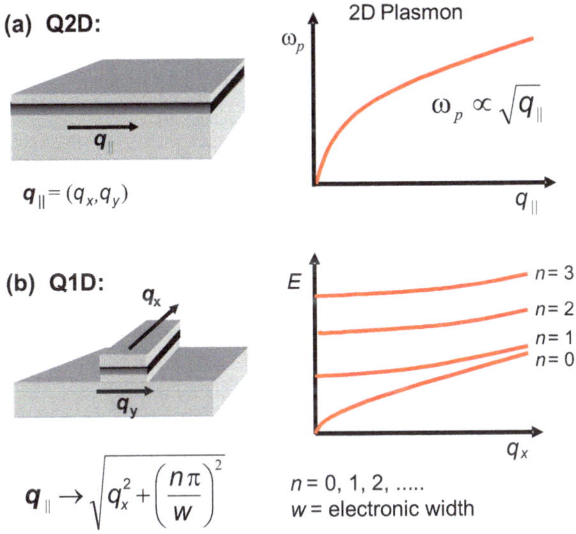

Fig. 6.3. (**a**) Schematic picture of a GaAs–AlGaAs quantum–well structure, hosting a Q2D electron system, and inplane wave–vector dispersion of the 2D plasmon. (**b**) Schematic picture of a deep–etched quantum wire, and confined plasmon dispersion. In a quantum wire, the wave vector $q_{\|}$ is quantized by the given relation in y direction and quasi–continuous in x direction

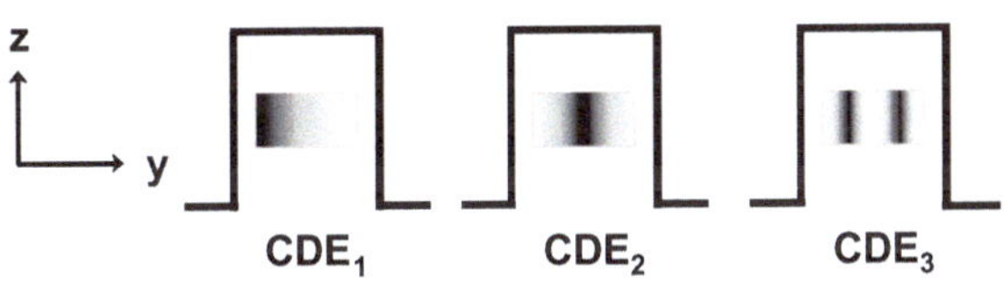

Fig. 6.4. Schematic snapshot picture of the induced charge densities for the first three confined plasmons, CDE_n with $n = 1, 2, 3$, in a quantum wire, perpendicular to the wire direction. A dark color should mean a large induced electron density

6.2.2 Experimental Spectra and Wave–Vector Dependence

Before discussing the microscopic origin of the confined modes, in this section experimental observations shall be presented to introduce the electronic excitations in quantum wires. In particular, in different resonance experiments it was possible to observe all types of electronic elementary excitations – SPE's, SDE's, and CDE's – in quantum wires (as well as in quantum dots) [3, 21, 22]. This allowed one, e.g., to do a detailed analysis of many–particle interactions in these low–dimensional electron systems [3, 21].

In Fig. 6.5, polarized and depolarized spectra of electronic excitations in a quantum–wire sample are shown. Several peaks can be observed, which can

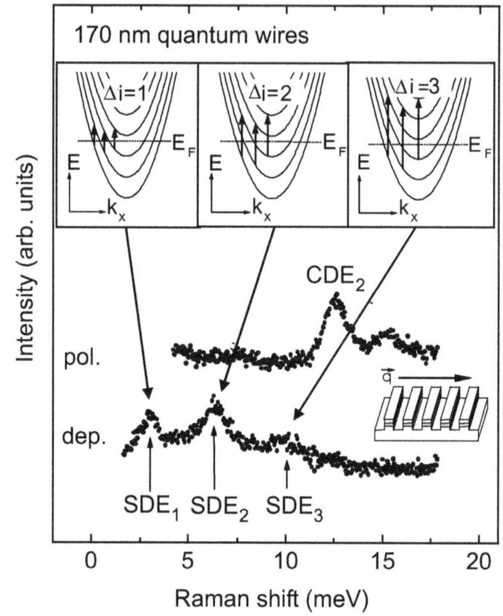

Fig. 6.5. Depolarized and polarized Raman spectra of electronic excitations in a quantum–wire sample at $T = 12$ K. The insets shows a schematic picture of the single–particle transitions, which predominantly contribute to the observed excitations. The wave–vector transfer q_y perpendicular to the wires was $q_y \approx 1.3 \times 10^5$ cm^{-1}

be identified as either SDE's or CDE's due to polarization selection rules. These spectra were recorded at a laser frequency well above the effective band gap of the underlaying 2D structure. From investigations at different laser frequencies [3] it is known that here the collective excitations are created by a third–order excitonic scattering mechanism as discussed in Sect. 4.2.1 for 2D intersubband excitations. At the end of this section we will see that also in the sample discussed here, at conditions of extreme resonance SPE's show up. The index i of the labels SDE_i or CDE_i in Fig. 6.5 gives the change in lateral quantum number for single–particle transitions, which dominantly contribute to the observed collective excitations[1]. We will discuss this in detail below in Sect. 6.3. The rather large differences in the energetic positions of the SDE_i and CDE_i are due to many–particle interactions [3]. As discussed above, in a macroscopic view, the CDE's in wires can be regarded as electron–density oscillations perpendicular to the wire direction with, e.g., one node in the induced density for the CDE_1, and two nodes for the CDE_2 (cf. Fig. 6.4). The peculiarity in wire and dot structures with parabolic bare potential is

[1] We note that in those samples there are about 9 occupied 1D subbands. For simplicity, in the insets smaller electron numbers are considered.

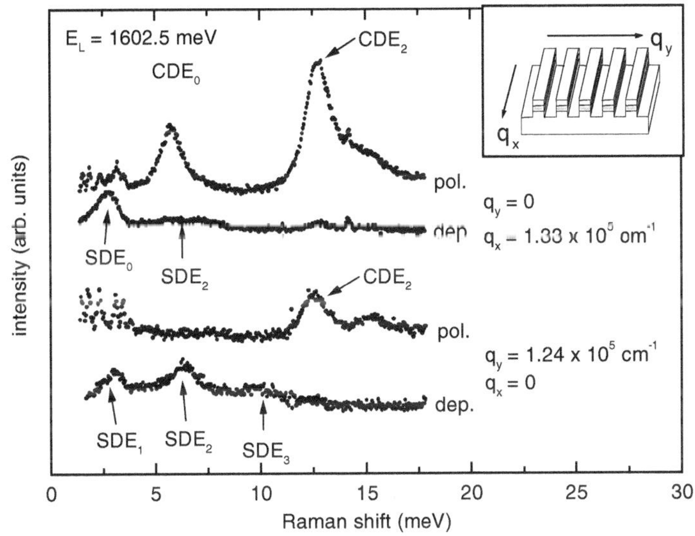

Fig. 6.6. Depolarized and polarized inelastic light scattering spectra of electronic excitations in a quantum–wire sample with 170 nm wire width at a temperature of $T = 12$ K. A wave vector q was transferred either parallel (*upper part*) or perpendicular (*lower part*) to the wire direction

that the energy of the CDE_1, the so called Kohn's mode, is exactly equal to the single–particle spacing of the bare potential. The reason is the so called generalized Kohn theorem [23, 24]: The CDE_1 is in the long–wavelength limit a rigid center–of–mass oscillation of all electrons. Therefore, this mode has a large dipole moment and can be observed in far–infrared transmission experiments [25]. It shall be emphasized here that the general parity selection rules for inelastic light scattering experiments, which we found in chapter 5 for experiments on quantum dots, can be found here again in quantum wires: The even parity plasmon mode, CDE_2, is the dominant mode in the polarized spectrum, and the dipole mode CDE_1 is not observable at all in Fig. 6.5. Again, as in the case of dots, these parity selection rules are much better fulfilled for CDE's than for SDE's.

By applying a finite $\boldsymbol{q} = (q_x, 0)$ *parallel* to the wires, also travelling charge–density waves *along* the wires are induced. Corresponding spectra are displayed in the upper part of Fig. 6.6. In the lower part, for comparison, the same spectra as in Fig. 6.5a, for a wave–vector transfer $\boldsymbol{q} = (0, q_y)$ *perpendicular* to the wires, are shown. The SDE_0 and CDE_0 are, respectively, the intraband spin-density and charge-density excitations, which can only be observed at finite q_x along the wires. The q_x dispersions of these modes are shown in Fig. 6.8, below. The SDE_1, SDE_2, and SDE_3 are in this situation ($q_x \neq 0$) subject to strong Landau damping. Therefore, in the spectra at

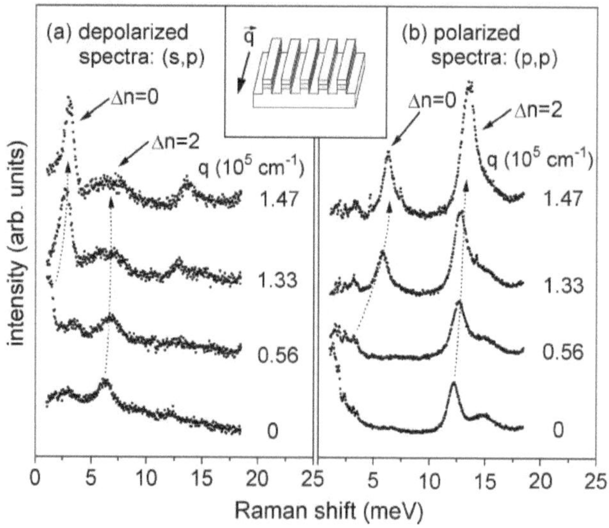

Fig. 6.7. Depolarized and polarized inelastic light scattering spectra of electronic excitations in a quantum–wire sample with 170 nm wire width for different wave–vector transfer q_x parallel to the wire direction [Reprinted with permission from [3]. Copyright (1996) by the American Physical Society]

$q_x = 1.33 \times 10^5$ cm^{-1} in Fig. 6.6, only the SDE$_2$, which is the strongest peak at $q_x = 0$, remains as a very broad feature. The evolution of the spectra with wave–vector transfer q_x, parallel to the wire direction, is shown in more detail in Fig. 6.7a for SDE's and in Fig. 6.7b for CDE's. In Fig. 6.7a it can be seen that the SDE's are subject to Landau damping, when they enter the single–particle continua. This behavior is discussed in more detail in [3].

In Fig. 6.8, the results discussed so far are summarized: The observed positions of SDE's, SPE's (see below), and CDE's are marked by open circles, full circles and full squares, respectively. Thin grey lines indicate the borders of the intersubband single–particle continua. In this sample, the number of occupied Q1D subbands was estimated to be about 9. The above discussed forbidden regions in the continua have been omitted in Fig. 6.8. One can see that the SDE's are only slightly shifted to lower energies, compared to the corresponding SPE's. Hence, already for a small wave–vector transfer q_x they enter the corresponding single–particle continua, which is manifested by Landau damping, i.e., a broadening of the excitations (Fig. 6.7a). On the other hand, there is a large depolarization shift of the CDE's with respect to the corresponding SPE's. Therefore, in the displayed wave–vector range, the CDE's do not enter the continua of single–particle transitions, from which they originate. As a result, they show no effect of Landau damping.

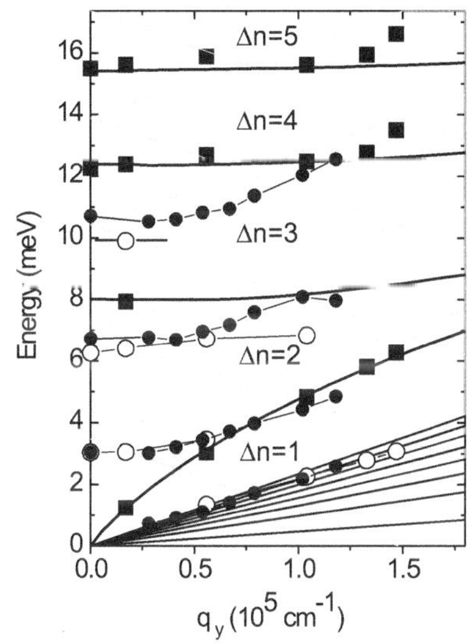

Fig. 6.8. Wave–vector dispersions of the observed excitations in a quantum–wire sample. SDE's are marked with open circles, SPE's with filled circles, and CDE's with filled squares. The grey lines mark the borders of the calculated single–particle continua

The resonances of collective and single–particle–like excitations occur at different laser energies. In the introductory Chap. 4 it is discussed that in quantum wells the intersubband SPE becomes very strong under conditions of extreme resonance, when the laser energy comes close to the fundamental bandgap. The collective excitations, on the other hand, show sharp resonance profiles at higher laser energies, due to a third order excitonic scattering process. We found a similar behavior in experiments on deep–etched dots (see Chap. 5) and on quantum wires. This can be deduced from Fig. 6.9. Figure 6.9 shows polarized spectra for different laser energies, E_L. It can be seen that, in the displayed experiment, the CDE_2 exhibits a very sharp (FWHM $\approx$ 3.5 meV) resonance profile with respect to the variation of the laser energy in a certain range of laser frequencies, which shows that this scattering by *collective* excitations is caused by the third–order TOP as discussed in Sect. 4.2. If the laser frequency is lowered, additional broad features appear, which then slowly evolve into intense peaks, by further lowering E_L towards the bandgap. These peaks then dominate the spectra and are present in both polarization configurations. Therefore, they have been interpreted within the framework described in Sect. 4.2 as SPE's, which are caused by a SOP under conditions of extreme resonance [3]. From the experimental spectra in Figs. 6.6 and 6.9 we can deduce for the corresponding collective SDE's and CDE's directly the energy renormalizations due to many–particle interactions. In a theoretical

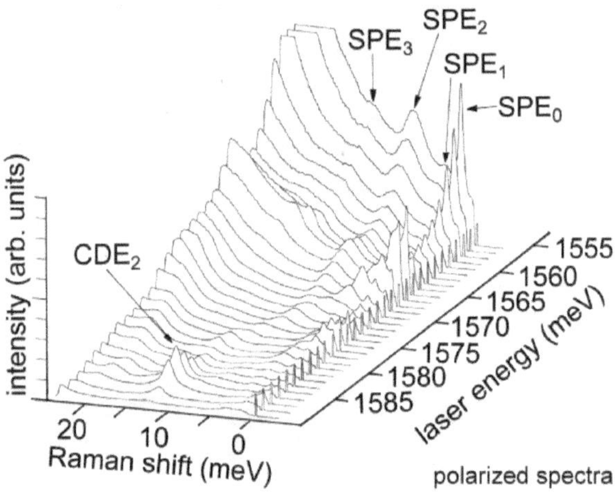

Fig. 6.9. Polarized inelastic light scattering spectra of electronic excitations in a quantum–wire sample for different laser energies. The background is due to hot luminescence

work in the framework of the time–dependent Hartree–Fock approximation and also the local–density approach, it was demonstrated that the finite width of the underlaying 2D structure has to be taken into account for a correct description of the energies of the SDE's in quantum wires [26].

6.3 Confined and Propagating 1D Plasmons in a Magnetic Field

In order to gain more insight into the microscopic structure of the observed excitations in quantum wires, we have also performed experiments in an external magnetic field B, which was oriented perpendicular to the plane of the wire array. Furthermore, we have compared our experimental results with self–consistent calculations of the dynamic Raman response. In the first part of this section we give a microscopic picture for confined plasmon modes in quantum wires [11]. As examples, in which complex way the internal structure of the excitations is influenced by a magnetic field, we discuss in the second part (i) the anticrossing of the CDE's in quantum wires due to the coupling with Bernstein modes [11, 12], which can in a similar manner also be observed in quantum dots, and (ii) 1D plasmons which represent skipping–orbit modes [12].

6.3.1 Microscopic Picture for Confined Plasmons

From our investigations in Sects. 6.2.2 and 5.3.3, we know that the energies of the SDE's in quantum wires and dots are very close to single–particle transitions. We have also emphasized there that the excitations can be characterized by specific changes in lateral quantum number. To proof this, self–consistent Hartree calculations of the 1D single–particle subband structure and RPA calculations of Raman scattering cross sections were performed. This enables a microscopic understanding of the internal structure of CDE's in quantum wires, which exhibit rather large collective shifts.

To demonstrate the complexity of the behavior of magnetoplasmons[2] in quantum wires, we start by displaying in Fig. 6.10 the experimental mode dispersions in a perpendicular magnetic field (full symbols). Rather different anticrossings occur, depending on the directions of wave–vector transfer (Figs. 6.10b and 6.10c). In the following we want to highlight the internal structure of these complicated hybrid excitations. We start our discussion by investigating the internal structure of the confined plasmon modes at $B = 0$ (points at $B = 0$ in Fig. 6.10c) and will later elucidate the behavior at finite magnetic fields.

The Raman spectra $I(q, \omega)$ were calculated from the imaginary part of the density–density correlation function $D(q, \omega)$ in RPA for nonresonant conditions [11]

$$I(q, \omega) \sim \mathrm{Im}\,\{D(q, \omega)\} \ . \tag{6.2}$$

As pointed out before, this will deliver us the correct mode positions but will not allow to draw conclusions about mode intensities. Figure 6.11 shows calculated and measured Raman spectra of confined plasmons in a quantum–wire sample with 170 nm geometrical wire width. In these experiments, a wave vector q was transferred perpendicular to the wire direction to effectively induce electron motion in the direction of the lateral confinement, i.e., perpendicular to the wires. In contrast to the polarized spectra, displayed in Figs. 6.5 and 6.6 in Sect. 6.2.2, three lines can be observed in Fig. 6.11. This is because the experimental spectrum in Fig. 6.11 was recorded at a different laser energy, where also the excitations with odd parity are resonantly enhanced. The three confined plasmons are the CDE_1, CDE_2, and CDE_3. Let us for a moment neglect Coulomb interaction and consider the 1D single–particle subbands, only. Then, we can expect single–particle transitions where the 1D subband quantum number i changes for example by 1, 2 or 3 (insets of Fig. 6.11) as we have already considered in Sect. 6.2.2. By comparison to our calculations of the subband structure, we find that the subband spacing at the Fermi level in the wire sample is about 3.5 meV. This is in agreement with magnetic depopulation measurements. Also inelastic light scattering experiments at lower excitation energy, which directly show SPE's, yield the same energy [3, 27] (see Fig. 6.9). In contrast, the energy of the first observed mode in Fig. 6.11 is about 8 meV. This large difference between the single–particle energy and the energy of the observed excitation is due to Coulomb interaction, which for 2D electron systems is called depolarization shift and which was investigated in Sect. 3.3.1. The single–particle transitions couple

[2] In an external magnetic field, CDE's are typically called *magnetoplasmons*.

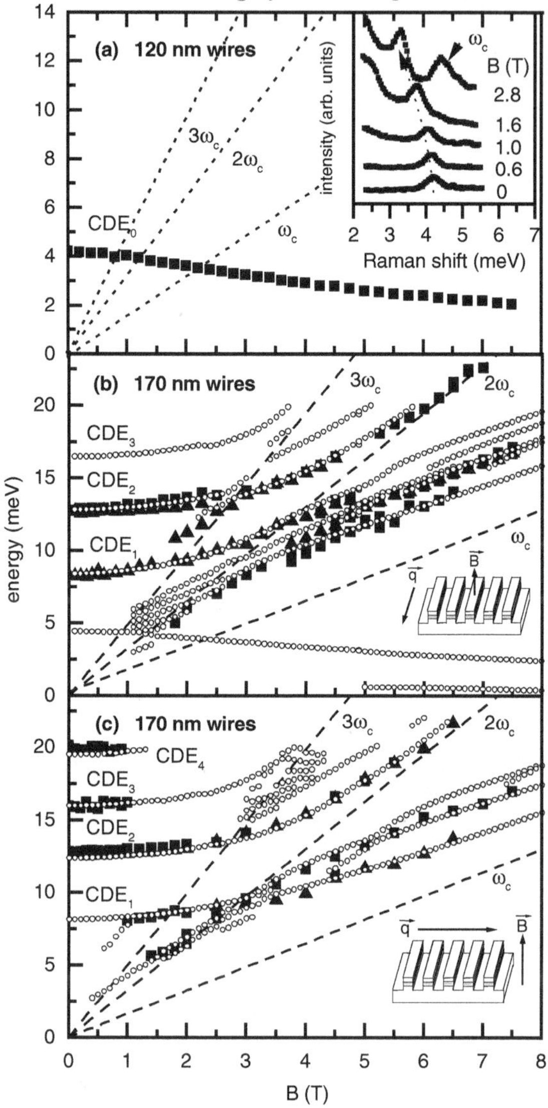

Fig. 6.10. Experimental (*full symbols*) and theoretical (*open symbols*) magnetic-field dispersions of magnetoplasmons in quantum-wire samples for different directions of wave-vector transfer q with respect to the wire direction

due to Coulomb interaction. The theoretical curve in Fig. 6.11 indicates that this many-body effect can be very well reproduced in the RPA calculations. Conceptional, within RPA, single-particle transitions with arbitrary Δi contribute to each excitation. However, the calculations show that, at $B = 0$ T, the first confined plasmon mode predominantly consists of single-particle transitions where the 1D subband quantum number i changes by 1 (see

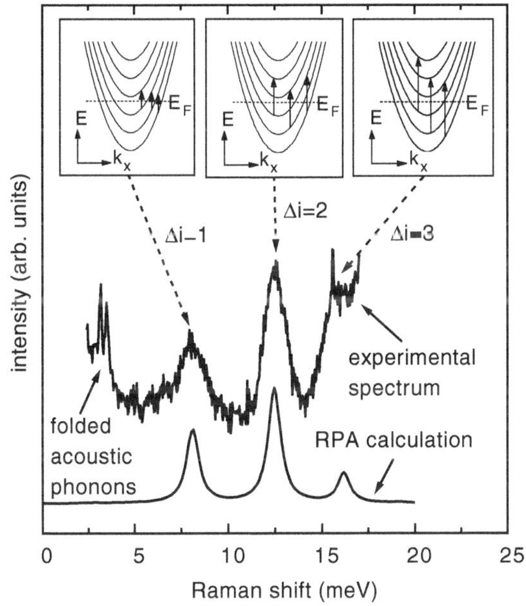

Fig. 6.11. Experimental and theoretical polarized Raman spectra of a quantum–wire sample with 170 nm wire width at $T = 12$ K. The wave vector has been transferred perpendicular to the wires. The insets show the single–particle transitions that predominantly contribute to the observed confined plasmon modes [Reprinted with permission from [11]. Copyright (1996) by the American Physical Society]

inset of Fig. 6.11). Correspondingly, for the second confined plasmon we have $\Delta i = 2$ and for the third $\Delta i = 3$. To demonstrate this on a microscopic level, we show in Fig. 6.12 calculations of the density distribution $\delta N(y)$, which is induced during the inelastic light scattering process for the first (a) and second (b) confined plasmon modes. Here, again y is the lateral direction perpendicular to the wires. The solid curves in Fig. 6.12a and 6.12b show the results if single–particle transitions with $\Delta i = 1, 2, 3, 4$, and 5 are included, whereas the dashed lines have been calculated for, respectively, $\Delta i = 1$ or $\Delta i = 2$ transitions, only. The dashed and solid curves do not differ significantly so that we can draw the conclusion that, e.g., the first confined plasmon mode predominantly consists of single–particle transitions with $\Delta i = 1$. Furthermore, Fig. 6.12a shows that for the first confined plasmon the density distribution is asymmetric with respect to the center of the wire at $y = 0$. This mode is in the long–wavelength limit the Kohn's mode, which has a large dipole moment and is therefore FIR active. For a parabolic confinement, this mode is the only one which is observed in direct FIR absorption. In contrast, the induced density of the second confined plasmon is symmetric (Fig. 6.12b) and it contains no dipole moment at all. This mode with even parity is most

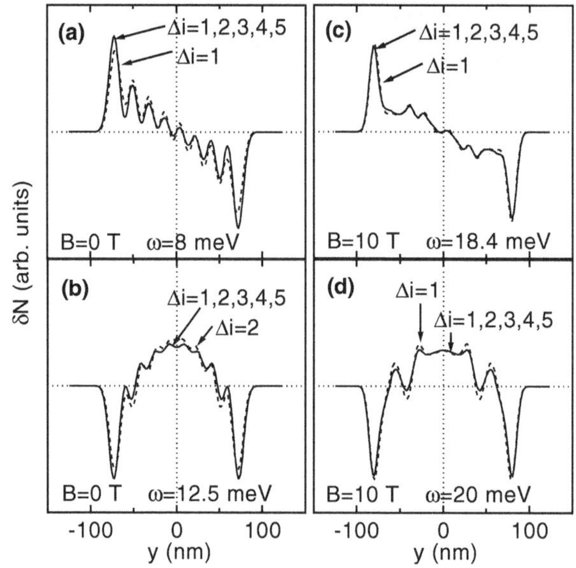

Fig. 6.12. Calculated density distributions in the inelastic light scattering experiments for the first (**a**) ($\Delta i = 1$) and second (**b**) ($\Delta i = 2$) confined plasmon modes in a wire sample with 170 nm geometrical width for $B = 0$ T. y is the lateral direction perpendicular to the wire direction. In (**c**) and (**d**), the corresponding distributions for the two lowest modes at $B = 10$ T are shown. Here both modes have $\Delta i = 1$ character [Reprinted with permission from [11]. Copyright (1996) by the American Physical Society]

prominent in the Raman spectra as can be seen in Fig. 6.11, and also, more clearly, in Fig. 6.5 in Sect. 6.2.2.

6.3.2 Coupling with Bernstein Modes

We will now turn to the behavior at finite magnetic field. Figure 6.10c shows the magnetic–field dispersions (solid symbols) of the confined plasmons for a wave–vector transfer perpendicular to the wires. The CDE_1, which is in the long–wavelength limit the Kohn's mode, shows only a very small splitting in the measured B range. This shows that the external lateral potential is to a good approximation parabolic because in this case the generalized Kohn's theorem holds. Interestingly, the second confined plasmon, the CDE_2, exhibits a strong anticrossing at $2\omega_c$ due to the coupling with Bernstein modes. The CDE_2 is a Raman–active mode with even parity and has no dipole moment at all (see also discussion in the previous section) [11]. This implies that it purely consists of relative motions of the interacting electrons, so that it is clear that here the generalized Kohn's theorem [24] can not be applied. The small open circles in Fig. 6.10 are results of a calculation.

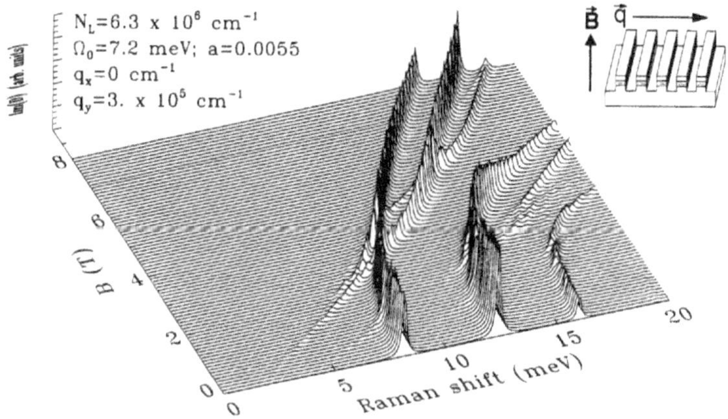

Fig. 6.13. Calculated inelastic light scattering spectra of confined magnetoplasmons in quantum wires. N_L is the linear density, Ω_0 the quantization energy of the external parabolic potential, q_x the wave–vector transfer parallel, and q_y the wave–vector transfer perpendicular to the wire direction

Figure 6.13 shows the corresponding theoretically determined inelastic light scattering spectra. Here, only a slight nonparabolicity $a = 0.0055$, according to $V(y) = 1/2\hbar\Omega_0[(y/l_0)^2 + a(y/l_0)^4]$, of the external potential $V(y)$ has been assumed. l_0 is the characteristic length. The theoretical curves displayed in Fig. 6.10 show the dispersions of the strongest modes. The calculations also demonstrate that in the anticrossing regime the modes can no longer be characterized, as for other magnetic fields, by specific single–particle transitions, but rather a strong intermixing of different single–particle contributions occur [11]. This conclusion can be drawn from Fig. 6.14. Here the calculated induced density is shown for the range where the small splitting of the CDE_1 occurs, i.e., at $B = 2.8$ T and $\omega \approx 9$ meV. Figures 6.14a and 6.14b display the induced densities for the lower branch ω_-, and Figs. 6.14c and 6.14d for the higher energy branch ω_+. It clearly demonstrates that in this anticrossing regime the modes can no longer be characterized by either $\Delta i = 1$ or $\Delta i = 2$ transitions but rather both types must be included in the calculations to reproduce the correct density distribution. On the other hand, at very high magnetic fields, far outside the anticrossing regime, the modes again have well defined character, as in the case $B = 0$. This can be seen from Figs. 6.12c and 6.12d in the previous section, where the calculated density distributions of the two lowest modes at $B = 10$ T are displayed.

In Fig. 6.10b, the results for a wave–vector transfer q in wire direction are displayed [12]. Here, the macroscopic density oscillations, which are characteristic for the excitations, consist of a traveling wave along the wires and, at the same time, an oscillation perpendicular to the wires. As we can see from Fig. 6.10b, in this case also the Kohn's mode exhibits a strong splitting in the

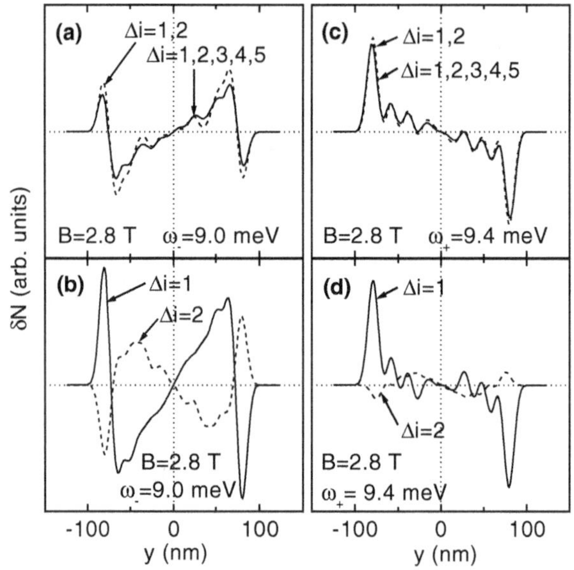

Fig. 6.14. Calculated density distributions at $B = 2.8$ T. (**a**) and (**b**) show the contributions to the lowest (ω_-), (**c**) and (**d**) to the higher (ω_+) of the split Kohn modes [Reprinted with permission from [11]. Copyright (1996) by the American Physical Society]

vicinity of $2\omega_c$, which demonstrates that Kohn's theorem does not hold, because we are no longer in the long–wavelength limit $q \approx 0$. Figure 6.15 shows the calculated spectra for this situation. The theoretical curves in Figs. 6.10b

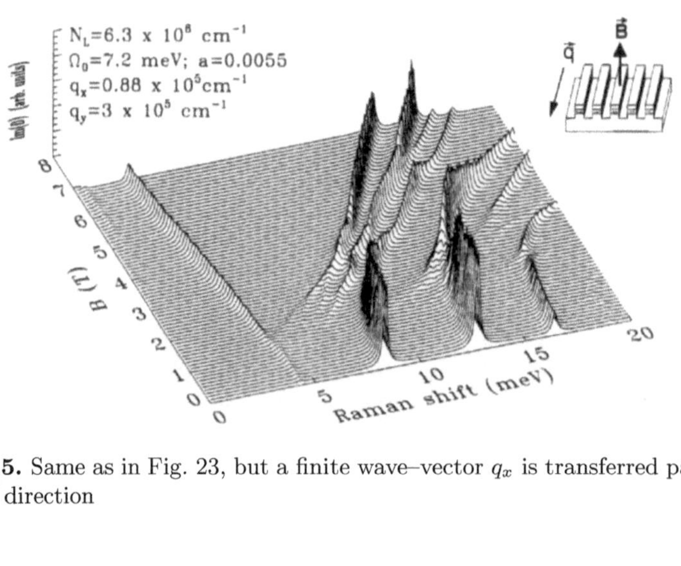

Fig. 6.15. Same as in Fig. 23, but a finite wave–vector q_x is transferred parallel to the wire direction

and 6.15 also show one mode, starting at about 4 meV at $B = 0$, which shows a negative B dispersion. This mode is in a microscopic picture an 1D *intra-band* plasmon (CDE_0), i.e., a plasma wave which propagates along the wires [28]. The negative B dispersion results from a skipping–orbit motion of the individual electrons at the edges of the wires. This mode could be detected in samples with relatively small periods and wire widths [12, 29]. For illustration, polarized spectra are displayed in the inset of Fig. 6.10a. We have also calculated the dispersion of this skipping–orbit mode (see, e.g., Fig. 6.15). The calculations, which are performed for isolated wires, reproduce the negative dispersion, however, they do not give the correct energy. The calculated energy is typically only half of the experimentally observed one. We attribute this experimentally observed energy increase to the coupling between wires in these small period arrays. Such an increase due to coupling is suggested by the calculation of Li and DasSarma [30]. Also the q dependence of the observed 1D plasmon at $B = 0$, which is plotted in Fig. 6.16, supports this interpretation. We note that these spectra were recorded under conditions of extreme resonance at a laser energy close to the effective bandgap where also SPE's can be observed. The experimental dispersion of the 1D plasmon has clearly a $\sqrt{q}$ dependence. For isolated wires, even in the case of up to about

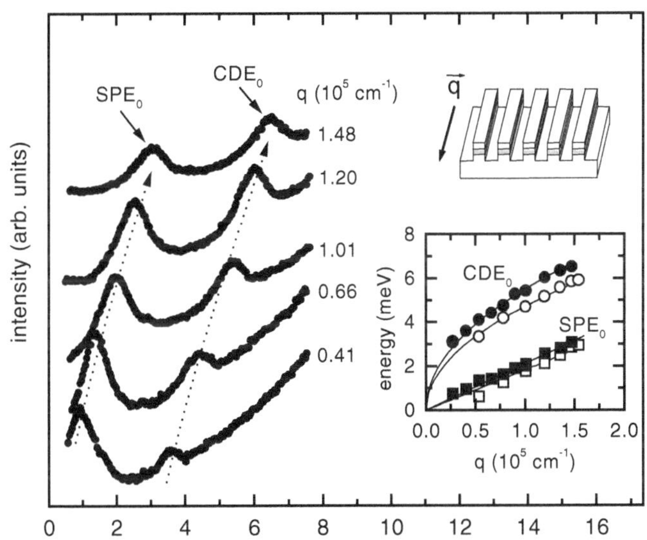

Fig. 6.16. Polarized inelastic light scattering spectra of the low–frequency 1D plasmon and SPE in a quantum–wire sample with 180 nm wire width and 360 nm period. The inset shows the q dispersions of the observed modes of a 180 nm (*solid symbols*) and a 150 nm (*open symbols*) wire sample [Reprinted with permission from [12]. Copyright (1997) by the American Physical Society]

four occupied subbands, which is the case for these narrow samples, our calculations, and also the experiment of Goñi et al. [5], yield a nearly linear q dispersion. Coupling between wires, however, leads to the experimentally observed $\sqrt{q}$ dependence, as calculated by Que [31].

In summary, in this section, a microscopic picture for the confined plasmon modes in quantum wires was given [11]. Furthermore, we have investigated the q and B dependence of magnetoplasmons in quantum wires. In quantum wires with nearly parabolic external potential, we find that for a wave–vector transfer *perpendicular* to the wires, higher index magnetoplasmons show strong coupling to Bernstein modes, whereas the Kohn's mode only weakly interacts [11, 12, 21]. On the other hand, if q is transferred *parallel* to the wires, also the Kohn's mode strongly splits in the vicinity of $2\omega_c$. Our experimental results are confirmed by self–consistent Hartree–RPA calculations. In short–period wire samples, an 1D plasmon, which shows a negative B dispersion due to a skipping–orbit motion of the individual electrons, could be observed. In these samples, evidence for a coupling between adjacent wires was found [12].

6.4 Towards the Tomonaga–Luttinger Liquid?

Much theoretical effort has been put into the description of one–dimensional electron systems in terms of a Luttinger liquid [15, 16]. To come experimentally as close as possible to a situation, which can prove the existence of such a correlated state, it is at least necessary to create Q1D systems in the quantum limit. However, the intrasubband CDE wave–vector dispersion, which is a typical quantity that can be extracted from inelastic light scattering experiments, is within some limit identical in the Luttinger liquid and the Fermi liquid theory [9, 17]. In Sect. 3.3 we have seen that in the RPA – which is a Fermi liquid theory – the energies of the collective excitations, i.e. plasmons, are determined by the poles of the wave–vector and frequency–dependent dielectric function, $\epsilon(\boldsymbol{q}, \omega)$. The dielectric function is given by the *exact* relationship [17]

$$\epsilon(\boldsymbol{q}, \omega) = 1 - V(q)\Pi(q, \omega) , \tag{6.3}$$

where $V(q)$ is the Fourier transform of the Coulomb interaction, and $\Pi(q, \omega)$ the exact irreducible polarizability function. The RPA is constructed on the approximation of replacing the exact $\Pi(q, \omega)$ by the irreducible polarizability, $\Pi_0(q, \omega)$, of a *noninteracting* electron gas, i.e.,

$$\epsilon_{\mathrm{RPA}}(\boldsymbol{q}, \omega) = 1 - V(q)\Pi_0(q, \omega) . \tag{6.4}$$

In a 1D electron gas, $\Pi_0(q, \omega)$ is given by [32]

$$\Pi_0(q, \omega) = \frac{m^*}{\pi q} ln \left(\frac{\omega^2 - \omega_-^2}{\omega^2 - \omega_+^2} \right) , \tag{6.5}$$

with

$$\omega_{\pm} = qv_F \pm \frac{\hbar q^2}{2m^*} , \tag{6.6}$$

where v_F is the 1D Fermi velocity. It can be shown [17] that from this one can get the long–wavelength plasmon dispersion

$$\omega_p = |q| \left(v_F^2 + \frac{2}{\pi} v_F V(q) \right)^{1/2} . \tag{6.7}$$

It is unique to the Q1D intrasubband plasmon dispersion (6.7) that it has the form of a depolarization–shifted SPE, since $\omega = qv_F$ is the dispersion of the intrasubband SPE (cf. Sect. 6.2.1 above). In all higher dimensions – Q2D and 3D – this is not the case. The important point, however, is that (6.7) is exactly the same as the eigenenergy of the elementary excitation in the Tomonaga–Luttinger model [33]. On the contrary, the Tomonaga–Luttinger model and the RPA strongly differ in the interpretation of the low–energy excitations: In the RPA one has intrasubband SPE's with a linear wave–vector dispersion and the above introduced intrasubband plasmon (CDE), as the collective excitation. The Tomonaga–Luttinger model predicts a perfect spin–charge separation, where SPE's do not exist. All elementary excitations are exclusively collective in nature: CDE's and SDE's. In the Tomonaga–Luttinger model, the interaction, V, is purely a parameter, while within RPA (or TDLDA) $V(q)$ is the matrix element of the $1/r$ Coulomb interaction in the lowest quantized subband. Therefore, it is exactly known, if the confinement potential is known. For Q1D systems it is well–known [32] that for $q << 1/a$, where a is the typical confinement (or electronic) width, the Fourier transform of the interaction potential, $V(q) \propto |ln(qa)|$. This leads in the long–wavelength limit to

$$\omega_p \propto |q||ln(qa)|^{1/2} , \tag{6.8}$$

which is for very small q values valid, only. This relation was experimentally verified for the first time in experiments of Goñi et al. [5] on GaAs–AlGaAs quantum wires with two occupied Q1D subbands. In 2000, Perez et al. [9] confirmed again this relation in experiments on deep–etched quantum wires, which were in the 1D quantum limit. Nevertheless, as already mentioned, these experiments do not allow one to distinguish between a Luttinger–liquid versus a Fermi–liquid behavior.

The wave–vector dependence of the Q1D intraband plasmon for very narrow quantum wires is obvious from Fig. 6.17. For these experiments, quantum wire samples with different geometrical widths between about $w = 250$ nm and $w = 50$ nm were prepared on 25 nm wide GaAs–AlGaAs single quantum wells by laser–interference lithography and RIE etching [34]. The observed mode positions are plotted in Fig. 6.18 versus the transferred wave vector, q_x, in wire direction. The solid lines were calculated following a model of Gold

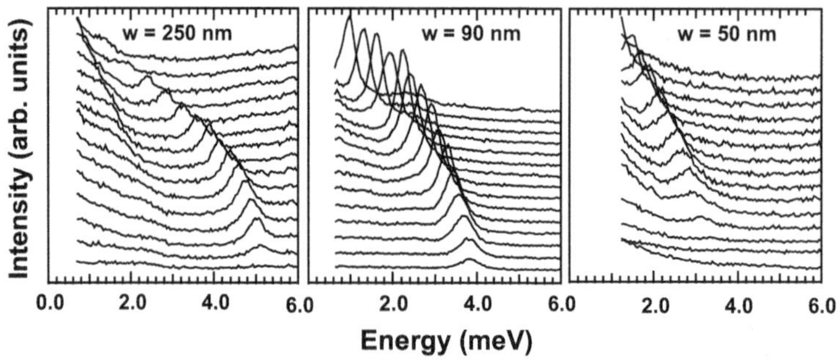

Fig. 6.17. Polarized inelastic light scattering spectra of the low–frequency 1D plasmon, CDE_0, in quantum–wire samples with 250 nm, 90 nm, and 50 nm geometrical wire widths. The transferred wave vector in wire direction, q_x, increases from 0.3×10^5 cm^{-1} to 1.6×10^5 cm^{-1} from top to bottom spectra

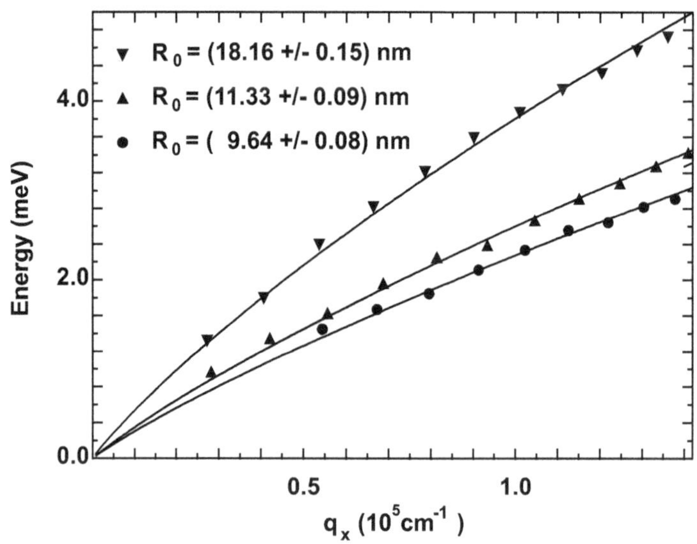

Fig. 6.18. Mode positions of the 1D intraband plasmon, CDE_0, from Fig. 6.17 versus transferred wave vector q_x. The *solid lines* are fits from a theoretical model (see text), which delivers the displayed parameters for the radius of the electronic wire

and Ghazali [35], which allows one to calculate the plasmon dispersion analytically. The model assumes a cylindrical shape of the Q1D electron system. With this assumption one is able to calculate the plasmon frequency to be

$$\omega_p = qR_0\Omega_0\sqrt{\left|ln\left(\frac{qR_0}{2}\right)\right|}\,, \tag{6.9}$$

with

$$\Omega_0^2 = \frac{N_{1D}e^2}{2\pi\epsilon\epsilon_0 m^* R_0^2} \,. \tag{6.10}$$

For the theoretical curves in Fig. 6.18, the electronic radius, R_0, and the electron density, N_{1D}, were used as fitting parameters. The numbers for R_0, displayed in Fig. 6.18, show that the estimates of the electronic widths, $2R_0$, from this model are much smaller than the geometrical wire widths w. This is however reasonable because of the depletion of the electronic system close to the surfaces. Clearly, the model of Gold and Ghazali, which assumes a wire in the quantum limit, is adequate for very narrow wires, only. For the wire with 50 nm geometrical width, the fit delivers an electronic diameter of $a = 2R_0 \approx 20$ nm and a 1D electron density of $N_{1D} = 6.5 \times 10^5$ cm^{-1}. With the relations $k_F = \frac{\pi}{2}N_{1D}$ and $E_F = \frac{\hbar^2}{2m^*}\frac{\pi^2}{4}N_{1D}^2$ for a 1D system, this density would yield a Fermi energy of $E_F \approx 48$ meV, which would clearly mean that the system is not in the Q1D quantum limit, since the 1D subband spacing is much below this value. How delicate such estimates are, can be seen by the fact that with the model of Eliasson et al. [20] (see Sect. 6.2) applied to the data of the narrowest wires in Fig. 6.18, almost the same electronic width of $a \approx 24$ nm can be estimated. However, the density in this case is about $N_{1D} = 1.7 \times 10^5$ cm^{-1}. For such a density, the wires with $w = 50$ nm geometrical width (narrowest wires in Fig. 6.18) would clearly be in the Q1D quantum limit, since a Fermi energy of $E_F \approx 3$ meV would result in this case. Apparently, the 1D electron density N_{1D} is a very critical parameter. Figure 6.19 displays the experimentally observed mode positions of the 1D intraband plasmon CDE$_0$ together with the intersubband plasmons, which were detectable in the experiments. The solid lines are the just mentioned fits by the model of Eliasson et al. [20]. This model delivers an electronic

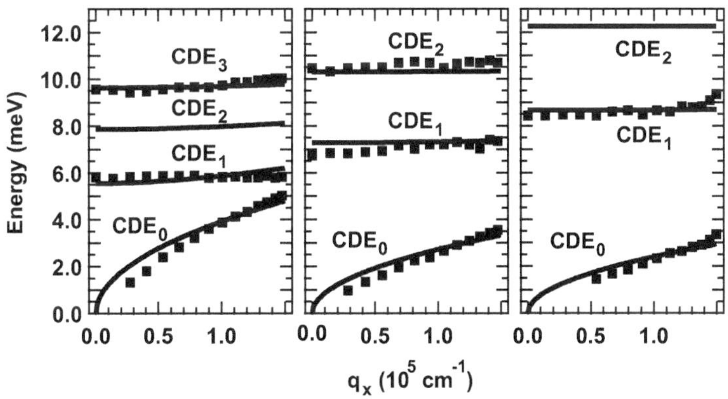

Fig. 6.19. Mode positions of 1D intraband and intersubband plasmons in quantum wires with w = 250 nm, 90 nm, and 50 nm geometrical widths (from *left to right panels*). The *solid lines* are fits due to the model of confined plasmons

width a for each sample from which N_{1D} can be estimated via the relation $N_{1D} = N_{2D}a$ using the known density N_{2D} of the quantum–well wafer, which was used as a starting material.

In any case, the data available so far on electronic excitations in quantum wires does not allow one to prove or disprove the Luttinger–liquid nature of the Q1D electron system. M. Sassetti and B. Kramer pointed out [16] that, employing the Tomonaga–Luttinger model, the intraband SPE_0, which can be observed in some experiments in the polarized spectra (cf., e.g., Sect. 6.3.2) are actually collective SDE_0. The authors predict that both, the resonance behavior and the temperature dependence of this mode should be strikingly different from those of the SDE_0 mode, which should occur in the depolarized configuration. It is understood that this different behavior in both polarizations is a consequence of the Luttinger–liquid nature of the Q1D electron system. Unfortunately, this can not be proven experimentally by the data presented above, since there, in the very narrow quantum–wire samples, the low–energy SDE's can not clearly be detected. So, concluding, the proof of the Luttinger–liquid nature of Q1D electron systems in narrow quantum wires still remains to be an experimental challenge.

References

1. J. S. Weiner, G. Danan, A. Pinczuk, J. Valladares, L. N. Pfeiffer, and K. W. West: Phys. Rev. Lett. **63**, 1641 (1989)
2. T. Egeler, G. Abstreiter, G. Weimann, T. Demel, D. Heitmann, P. Grambow, and W. Schlapp: Phys. Rev. Lett. **65**, 1804 (1990)
3. C. Schüller, G. Biese, K. Keller, C. Steinebach, D. Heitmann, P. Grambow, and K. Eberl: Phys. Rev. B **54**, R17304 (1996)
4. G. Biese, C. Schüller, K. Keller, C. Steinebach, D. Heitmann, P. Grambow, and K. Eberl: Phys. Rev. B **53**, 9565 (1996)
5. A. R. Goñi, A. Pinczuk, J. S. Weiner, J. S. Calleja, B. S. Dennis, L. N. Pfeiffer, and K. W. West: Phys. Rev. Lett. **67**, 3298 (1991)
6. A. Schmeller, A. R. Goñi, A. Pinczuk, J. S. Weiner, J. S. Calleja, B. S. Dennis, L. N. Pfeiffer, and K. W. West: Phys. Rev. B **49**, 14778 (1994)
7. C. Dahl, B. Jusserand, and B. Etienne: Phys. Rev. B **51**, 17211 (1995)
8. F. Perez, B. Jusserand, and B. Etienne: Phys. Rev. B **60**, 13310 (1999)
9. F. Perez, B. Jusserand, and B. Etienne: Physica E **7**, 521 (2000)
10. A. R. Goñi, A. Pinczuk, J. S. Weiner, B. S. Dennis, L. N. Pfeiffer, and K. W. West: Phys. Rev. Lett. **67**, 1151 (1993)
11. C. Steinebach, R. Krahne, G. Biese, C. Schüller, D. Heitmann, and K. Eberl: Phys. Rev. B **54**, R14281 (1996)
12. E. Ulrichs, G. Biese, C. Steinebach, C. Schüller, and D. Heitmann: Phys. Rev. B **56**, R12760 (1997)
13. S. Reich, C. Thomsen, and J. Maultzsch: *Carbon Nanotubes, Basic Concepts and Physical Properties* (Wiley VCH, 2004)
14. W. Wegscheider, G. Schedelbeck, M. Bichler, and G. Abstreiter: in Festkörperprobleme / Advances in Solid State Physics, Ed. B. Kramer (Vieweg Braunschweig Wiesbaden 1999) Vol. 38, p. 153

15. S. Tomonaga: Prog. Theor. Phys. **5**, 544 (1950); J. M. Luttinger: J. Math. Phys. **4**, 1154 (1963); H. J. Schulz: Phys. Rev. Lett. **71**, 1864 (1993)
16. M. Sassetti and B. Kramer: Phys. Rev. Lett. **80**, 1485 (1998)
17. Q. P. Li and S. Das Sarma: Phys. Rev. B **45**, 13713 (1992)
18. D. W. Wang, A. J. Mills, and S. Das Sarma: Phys. Rev. Lett. **85**, 4570 (2000)
19. U. Merkt: Physica B **189**, 165 (1993)
20. G. Eliasson, J.-W. Wu, P. Hawrylak, and J. J. Quinn: Solid State Commun. **60**, 41 (1986)
21. Christian Schüller, in Festkörperprobleme / Advances in Solid State Physics, Ed. B. Kramer (Vieweg Braunschweig Wiesbaden 1999) Vol. 38, p. 167
22. Christian Schüller: Physica E **3**, 121 (1998)
23. L. Brey, N. Johnson, and B. Halperin: Phys. Rev. B **40**, 10647 (1989)
24. P. Maksym and T. Chakraborty: Phys. Rev. Lett. **65**, 108 (1990)
25. see D. Heitmann and J.-P. Kotthaus in *Physics Today*, June 1993, p. 56, and references therein
26. C. Steinebach, C. Schüller, G. Biese, and D. Heitmann: Phys. Rev. B **57**, 1703 (1998)
27. G. Biese, C. Schüller, T. Kurth, D. Heitmann, P. Grambow, and K. Eberl: Surf. Sci. **361/362**, 797 (1996)
28. T. Demel, D. Heitmann, P. Grambow, and K. Ploog: Phys. Rev. Lett. **66**, 2657 (1991)
29. E. Ulrichs, G. Biese, C. Steinebach, C. Schüller, D. Heitmann, and K. Eberl: Phys. Stat. Sol. **164**, 277 (1997)
30. Q. P. Li and S. DasSarma: Phys. Rev. B **43**, 11768 (1991)
31. Weiming Que: Phys. Rev. B **43**, 7127 (1991)
32. S. Das Sarma and W. Y. Lai: Phys. Rev. B **32**, 1401 (1985)
33. D. C. Mattis and E. H. Lieb: J. Math. Phys. **6**, 375 (1965); A. Luther and I. Peschel: Phys. Rev. B **9**, 2911 (1974); F. D. M. Haldane: J. Phys. C **14**, 2585 (1981)
34. Annelene Dethlefsen: Diplomarbeit, University of Hamburg, 2004
35. A. Gold and A. Ghazali: Phys. Rev. B **41**, 7626 (1990)

7 Tunneling–Coupled Systems

7.1 Introduction

The plasmon spectrum of spatially separated two–component plasmas without tunneling has been studied for quite some time (see, e.g., [1]). For finite Coulomb coupling between the layers, the intrasubband charge–density excitation spectrum consists of two modes: The optical plasmon (OP) where both layers oscillate in phase parallel to the layers (see Fig. 7.1a), and, the acoustic plasmon (AP) where the carriers in both layers oscillate out of phase (see Fig. 7.1b). At long wavelengths, the energy of the OP is proportional to $\sqrt{q}$ and the energy of the AP goes linear in q, where q is the wave vector parallel to the layers. It was shown [1] that at large spatial separation of the two layers, the AP can move outside of the continua of possible intraband single–particle transitions. The first experimental observation of coupled–layer plasmons by inelastic light scattering was reported by Fasol et al. [2] on GaAs-AlGaAs samples containing five layers in parallel. In Coulomb–coupled double quantum wells, the observation of AP and OP was reported by Kainth et al. [3].

During the past decade there has been a growing interest in *tunneling-coupled* bilayer systems. In those systems, the interplay between many–particle Coulomb interaction and tunneling coupling – which, in the first place, is a single–particle effect – can be nicely studied. The additional degree of freedom, which comes into play in these systems due to the tunneling coupling, is commonly known as the so called pseudo spin. For bilayers with a perfectly symmetric potential in growth direction, the ground state is determined by the tunneling–split symmetric and antisymmetric single–particle states. New phenomena are expected due to intra– and interlayer Coulomb interactions. In the past decade, quite a number of experimental [4, 5, 6, 7, 8, 9, 10, 11] and theoretical [12, 13, 14, 15, 16, 17, 18, 19, 20] papers appeared, concerned with tunneling–coupled systems. Most inelastic light scattering experiments so far have been performed in symmetric tunneling–coupled bilayer systems [6, 7]. For this specific case, theory predicts two excitations, an *optical* intraband plasmon, which evolves from the OP of a purely Coulomb–coupled system, and an intersubband plasmon (ISP) [17, 18], which is qualitatively new, when compared to the situation without tunneling. As in the case of a single layer, the intraband plasmon is an ex-

Christian Schüller: *Inelastic Light Scattering of Semiconductor Nanostructures*
STMP **219**, 145–159 (2006)
DOI 10.1007/3-540-36526-5_7

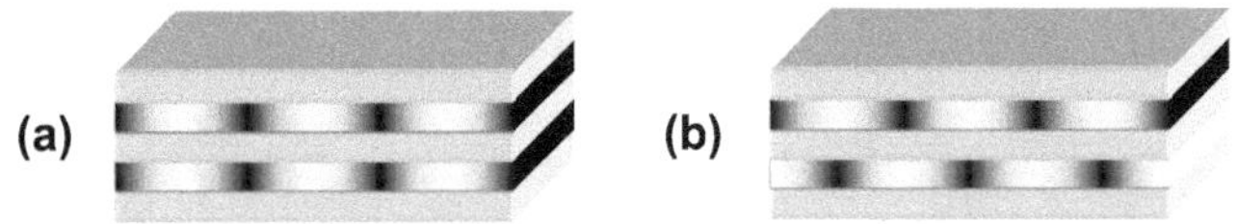

Fig. 7.1. Schematic snapshot picture of the electron–density distributions for (**a**) the optical plasmon (OP), and, (**b**) the acoustic plasmon (AP) of a Coulomb–coupled double–layer system. High electron density is indicated by a dark color

citation where, macroscopically, the total electron density in both layers is oscillating within the plane. The ISP, on the other hand, consists of transitions of electrons between the tunneling–split subbands, i.e., it is a density oscillation *between* the two layers. These excitations have been studied, e.g., in [6, 7], where in [6] it was reported that for occupation of both tunneling–split subbands, the exchange–interaction contributions from both subbands approximately cancel, and the intersubband SDE therefore has the same energy as the intersubband SPE. In this chapter we will concentrate on the spectrum of CDE's in tunneling–coupled systems to study the interplay between direct Coulomb and tunneling coupling. It can be shown that the energy of the AP is close to zero in the case of strong tunneling [20]. As soon as the potential deviates from perfect symmetry, however, the excitation spectrum becomes more subtle. In particular, the excitation energy of the AP increases again so that it should be observable in experiment.

In this chapter recent experiments on strongly coupled double quantum wells will be discussed. In those experiments, the theoretically predicted [20] plasmon spectrum could be observed, employing samples with external gates, which allow one to tune the symmetry of the double quantum–well potential in growth direction [11]. The chapter is finalized by a section, where first experiments on strongly tunneling–coupled quantum wires will be presented.

7.2 Charge–Density Excitation Spectrum in Tunneling–Coupled Double Quantum Wells

Figure 7.2 shows schematic pictures of the potentials and wavefunctions of double quantum wells in growth direction. In Fig. 7.2a, a symmetric potential is shown. The wavefunction in this case is perfectly symmetric or antisymmetric for the tunneling–split state with lower or higher energy, respectively. For a symmetric potential, the tunneling gap, Δ_{SAS}, is minimal. If the potential becomes nonsymmetric (Fig. 7.2b), the tunneling gap is larger than for the symmetric case, i.e., $\Delta > \Delta_{\mathrm{SAS}}$, and the wavefunctions have no longer a defined symmetry. With increasing tilt of the potential, the wavefunctions of

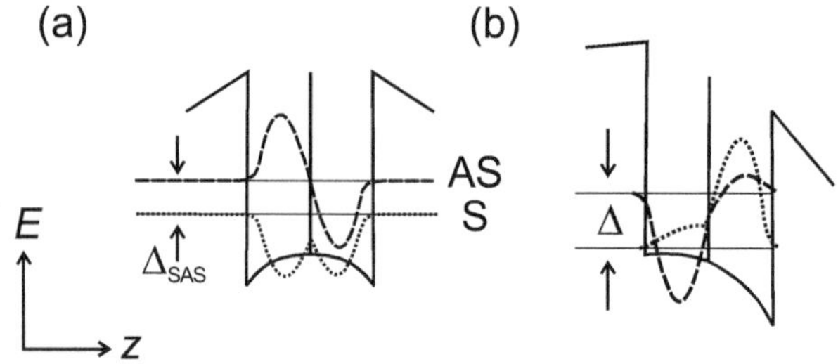

Fig. 7.2. (**a**) Schematic picture of a modulation–doped double quantum well with a thin tunneling barrier. The two lowest–energy tunneling–split states are indicated. The lowest–energy subband has a symmetric wavefunction (S, *dotted line*), while the wavefunction of the higher state is antisymmetric (AS, *dashed line*). (**b**) Same as (a) for a nonsymmetric double quantum–well potential

the two states become more and more localized in opposite wells. In [20], the symmetry of the potential is defined by a parameter $\sin(\Phi)$, with

$$\sin(\Phi) = \frac{\Delta_{\mathrm{SAS}}}{\Delta} . \tag{7.1}$$

It was shown [20] that in a tunneling–coupled bilayer system the low–energy CDE's can in a distinct way be influenced by the symmetry of the bilayer structure: In the general case of two–subband occupation, three low–energy CDE's exist, two intrasubband plasmons (AP and OP) and an intersubband plasmon (ISP), which originates from intersubband transitions between the tunneling–split ground–state subbands. In the following section, experimental investigations of the low–energy CDE's in modulation–doped GaAs–AlGaAs double quantum wells will be discussed [11]. In those experiments, semi-transparent gates allowed one to tune both, the carrier density and spatial symmetry of the double quantum well. By tuning the double–quantum–well potential from a symmetric to an asymmetric shape, it was possible to detect the AP of the tunneling–coupled system. Furthermore, an intriguing behavior of the ISP was found. It exhibits a crossover from a direct to an indirect excitation of the double quantum well. This crossover is expected to take place when the asymmetry of the potential is strong enough that, essentially, the wavefunction of the lower subband is localized in one well, and, the wavefunction of the upper subband is localized in the other well [11].

Before discussing the experiments, we will start in this section by investigating the spectrum of CDE's theoretically [21], following the considerations in [20]. There, the spectrum of CDE's is calculated in the framework of RPA, considering two delta layers, and incorporating the tunneling coupling by a coupling parameter $\sin(\Phi)$, as introduced above.

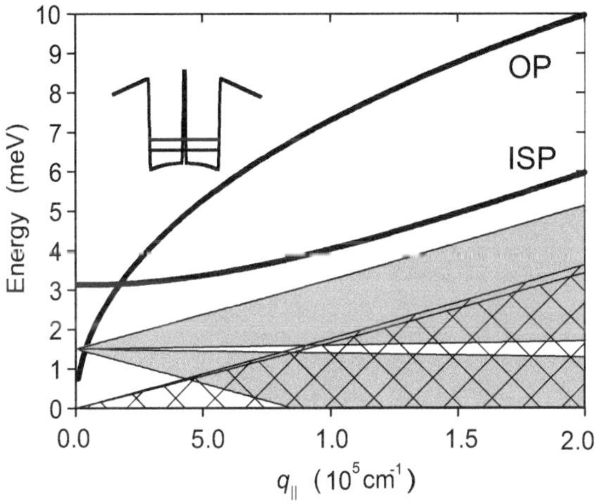

Fig. 7.3. Inplane wave–vector dispersions of electronic excitations in a tunneling–coupled double quantum well with symmetric potential. The *thick solid lines* indicate the dispersions of the ISP and OP. The hatched areas mark the single–particle continua of intraband excitations of the two tunneling–split subbands, and the *grey–shaded areas* those of intersubband transitions. Parameters are: separation of two layers, $d = 16$ nm, total carrier density, $n = 1.8 \times 10^{11}$ cm^{-2}, $\Delta_{\mathrm{SAS}} = 1.5$ meV

Figure 7.3 displays the calculated wave–vector dispersions of the low–energy CDE's (thick solid lines) of a perfectly symmetric double well together with the relevant intersubband and intrasubband single–particle continua. The tunneling gap was chosen to be $\Delta = \Delta_0 = 1.5$ meV, i.e., $\sin\Phi = 1$. As already discussed, for the symmetric case, the energy of the AP is close to zero and there are the ISP and the OP as the relevant collective modes. The density oscillations in the planes and perpendicular to the planes of the quantum wells, i.e., between the wells, are independent of each others and there is no coupling of the modes. This changes for nonsymmetric potentials, as shown in Fig. 7.4. Here, the coupling of ISP and OP manifests by an anticrossing of the two modes. Furthermore, for a nonsymmetric potential, the AP appears at finite energy within the intrasubband single–particle continua. The systematic behavior of the collective modes with symmetry of the double–quantum–well potential is displayed in Fig. 7.5. There, a fixed wave–vector transfer of $q = 0.84 \times 10^5$ cm^{-1} was chosen. The energy of the OP is nearly insensitive to the potential shape. It essentially depends on the total carrier density, n, which is kept fixed for the calculations displayed in Figs. 7.3 through 7.5. Of course, the quantized energy levels of the tunneling–split lowest–energy subband depend strongly on the symmetry of the potential. Hence, both, the energy of the ISP and the intersubband single–particle transitions depend strongly on $\sin\Phi$. With increasing asymmetry, the energy

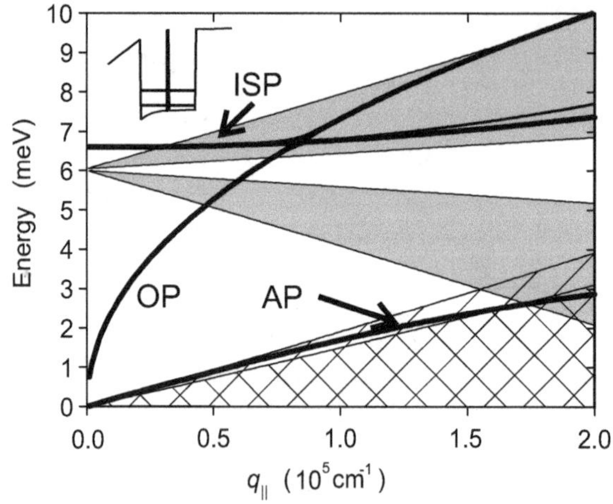

Fig. 7.4. Inplane dispersions of electronic excitations as in Fig. 7.3 but for $\sin\Phi = 0.25$, i.e., an asymmetric double–quantum–well potential

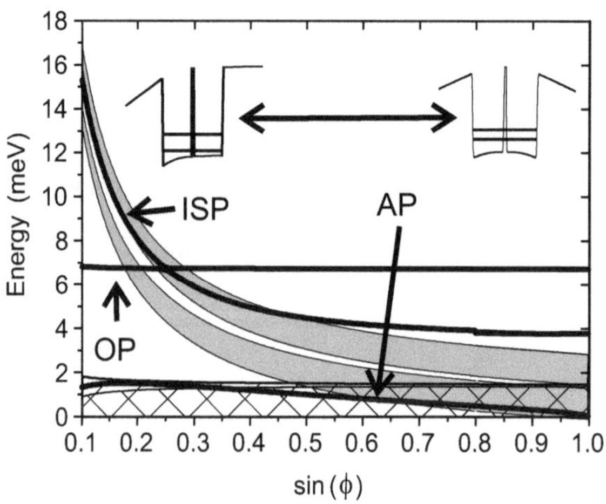

Fig. 7.5. Energies of electronic excitations of a double quantum well in dependence on the symmetry of the potential. The wave–vector transfer was kept fixed at $q = 0.84 \times 10^5\ \mathrm{cm}^{-1}$

of the AP moves towards the upper edge of the intrasubband single–particle continua.

In the following section, experiments on a modulation–doped GaAs–AlGaAs double quantum well, where n and $\sin\Phi$ could be controlled by external metallic gates, will be discussed.

7.3 Experiments on Tunable GaAs–AlGaAs Double Quantum Wells

The experiments, which will be discussed in this section, were performed on modulation–doped GaAs–$Al_{0.33}Ga_{0.67}As$ double quantum wells [11]. The samples consist of two 15 nm–wide GaAs quantum wells, separated by a 1 nm AlGaAs tunnel barrier. The AlGaAs barriers on both sides of the double quantum well are modulation doped, using Si delta layers. In the top barrier layer, two delta–doping layers were grown, separated by 28 nm AlGaAs and a 20 nm spacer layer to the upper GaAs well. In the lower barrier, one delta layer, separated by a 41 nm spacer from the lower GaAs well, was grown. Semitransparent Titanium gates were deposited on top of the samples. By applying a bias between the double quantum well and the gate, the carrier density and the self–consistent potential of the double quantum well in growth direction can be tuned.

In Fig. 7.6, polarized spectra for different inplane wave–vector transfer, q, are shown. Two lines, labelled as AP and OP, can be identified, which depend strongly on q. The broader mode, ISP, is nearly independent of q. Figure 7.7

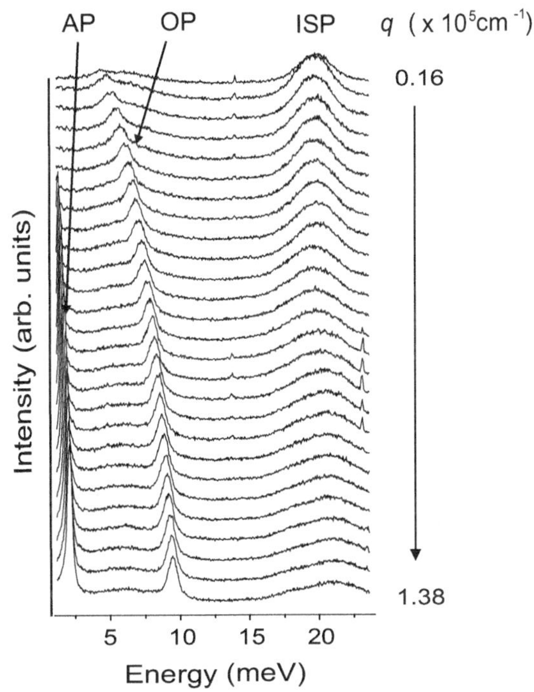

Fig. 7.6. Polarized spectra of a double quantum well for different inplane wave–vector transfer q

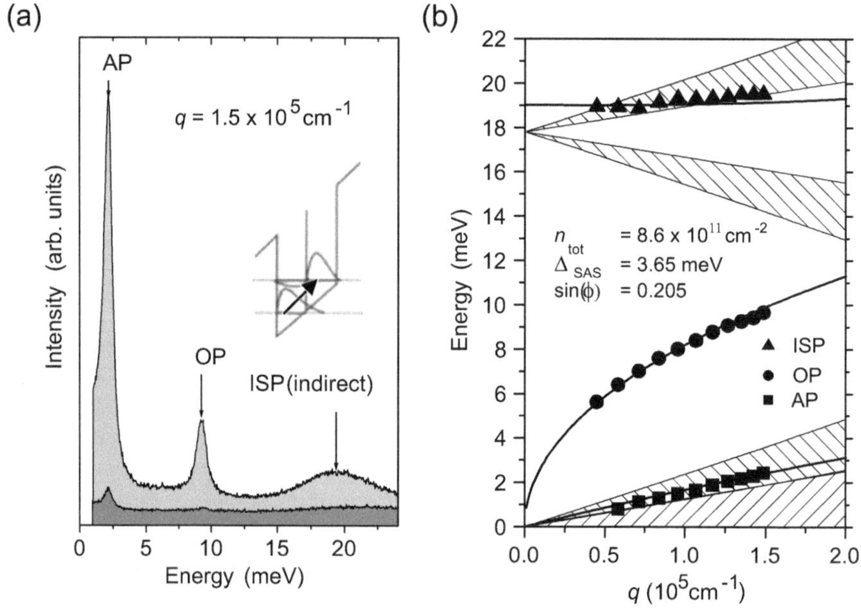

Fig. 7.7. (**a**) Polarized (*light gray*) and depolarized (*gray*) spectra of electronic excitations in an asymmetric tunneling-coupled GaAs-AlGaAs DQW. The inset shows a sketch of the DQW potential and wavefunctions. (**b**) Experimentally determined mode positions of the low-energy CDE's in dependence on the wave-vector transfer q parallel to the DQW for the same sample as shown in Fig. 1. The hatched regions mark the continua of intra- and intersubband single-particle transitions

displays a comparison of polarized and depolarized spectra of the low–energy excitations for large wave–vector transfer $q = 1.35 \times 10^5$ cm^{-1}. Since the three peaks, indicated in Fig. 7.7, appear dominantly only in the polarized spectrum, we can identify them as CDE's and rule out single–particle excitations. Remarkably, the excitation at 19 meV is much broader than the excitations at 2 meV and 9 meV, which have very similar linewidths. From considerations which will be discussed below, we find that the potential of the double–quantum–well structure in this experiment was strongly asymmetric with respect to the tunneling barrier (see inset of Fig. 7.7a). The interpretation of the CDE's displayed in Figs. 7.6 and 7.7a follows from the measured wave–vector dispersion, which is plotted in Fig. 7.7b (symbols). The lowest energy excitation (full squares in Fig. 7.7b) shows a linear q dependence and is therefore interpreted as the AP. The OP exhibits a square root–like behavior (full circles in Fig. 7.7b), and the highest energy excitation, the ISP between the tunneling–split subbands, depends only weakly on q (open symbols in Fig. 7.7b). The thick solid lines in Fig. 7.7b are calculated within RPA (for details of the calculation see [20]). For the sample displayed

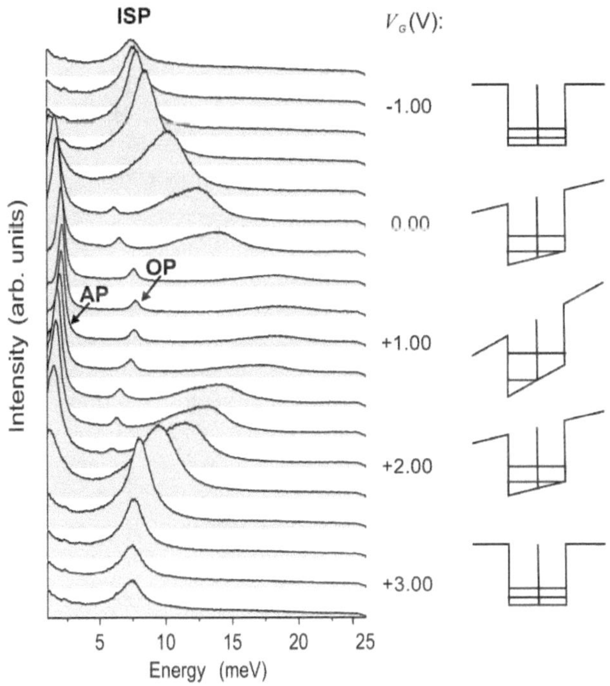

Fig. 7.8. Polarized Raman spectra for different gate voltages V (in steps of 0.25 V), applied between the DQW and a front gate. The wave-vector transfer is fixed at $q = 1.35 \times 10^5$ cm^{-1}. On the *right-hand side*, the shape of the DQW potential is sketched [Reprinted with permission from [11]. Copyright (2003) by the American Physical Society]

in Fig. 7.7b, the parameters are $n_{tot} = 8.6 \times 10^{11}$ cm^{-2} for the total carrier density, $\Delta_{SAS} = 3.65$ meV, and $\sin(\Phi) = 0.205$, which means that the double–quantum–well potential is significantly asymmetric. The theoretical considerations of the previous section showed that, in a tunneling–coupled bilayer, the AP can have a finite energy, and hence should be observable in experiment, for an *asymmetric* double–quantum–well potential, only. For the symmetric case, its energy tends to zero (is exactly zero for delta layers).

Figure 7.8 shows a series of polarized Raman spectra for different gate voltages V between -1.25 V and $+3.25$ V. At $V = -1.25$ V, the energy of the ISP is minimal (≈ 7.5 meV) and there is no AP (lowest–energy excitations in Fig. 7.8) visible at that gate voltage. From both it can be concluded that the potential is symmetric in this gate–voltage range. By tuning V towards positive values, the carrier density in the double quantum well increases, and, at the same time, the potential becomes more and more asymmetric (see schematic pictures on the right–hand side of Fig. 7.8). The energy of the OP is determined dominantly by the total carrier density of the double quantum well. Since in the experiment displayed in Fig. 7.8 the wave vector q is fixed, the energy of the OP is essentially proportional to $\sqrt{n_{tot}}$. In that sense, the OP serves as a direct monitor of the electron density in the experiment. With increasing carrier density and potential asymmetry (between $V = -1.25$ V and $V = 0.75$ V in Fig. 7.8), the energy and the linewidth of the ISP increases rapidly. At about $V = 0.5$ V, its intensity drops down within a relatively small voltage range. This was interpreted [11] as a crossover of the ISP from a direct excitation of the double–quantum–well structure to an indirect excitation: For a strongly asymmetric potential, the wave function of the lowest subband is located dominantly in one well (e.g., the left one, as sketched in the inset of Fig. 7.7a), and the wave function of the second subband is located in the other well. In this special situation, for an intersubband excitation, the electrons have to tunnel between the two layers. The strongly reduced overlap of the wave functions leads to the reduced intensity of the ISP.

In conclusion, in this section low–energy CDE's in tunneling–coupled double quantum wells were investigated in dependence of the carrier density and symmetry of the double quantum well. For asymmetric potentials, the AP of the tunneling–coupled system at low energies was observed and a crossover of the ISP from a direct to an indirect excitation of the coupled bilayer system was found.

7.4 Vertically–Coupled Quantum Wires

Now, the interesting question arises how the energies of the modes, which propagate as electron–density waves in the tunnelling–coupled 2DES, are altered if the double–quantum–well structure is patterned to quasi one–dimensional quantum wires. In the lateral direction perpendicular to the wires, we should expect a quantization of the intraband OP as well as of the AP modes. In this section we discuss the experimental observation of such confined modes in quantum wires, which are tunnelling–coupled in the vertical direction, i.e., the growth direction of the double–quantum–well structure.

Up to now there are no measurements reported on double–layered 1DES with *strong* coupling. Demel et al. [22] observed plasmon modes in weakly coupled double–layered 1DES by far–infrared spectroscopy. Theories describing 1DES with weak coupling can be found, e.g., in [23, 24, 25]. Until now there are also no theoretical publications concerning strong coupling in double–layered 1DES.

The investigated samples are modulation–doped GaAs–$Al_xGa_{1-x}As$ double quantum wells. They consist of two 15 nm–wide GaAs quantum wells, separated by a 1 nm AlGaAs tunnelling barrier. The AlGaAs barriers on both sides of the double quantum well are modulation doped using Si delta layers. For further reduction of the dimensionality, quantum–wire structures were prepared by holographic lithography and deep–reactive–ion etching (see Sect. 2.5), i.e., etching all the way through the double quantum well structure.

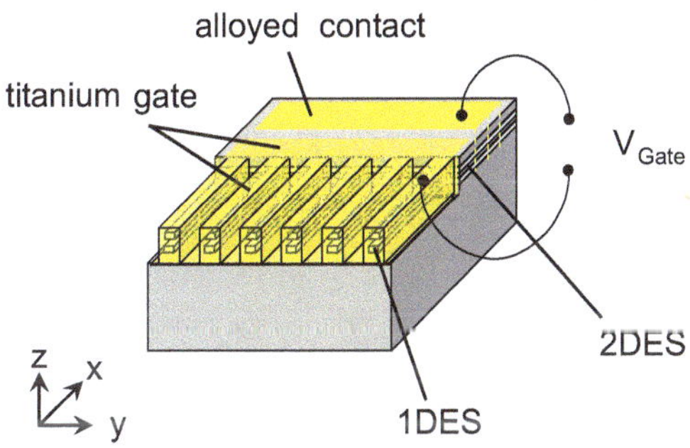

Fig. 7.9. Schematic picture of the investigated sample

Figure 7.9 shows a schematic picture of the prepared sample. It consists of two different lateral areas: a 1DES and a 2DES which allows one the measurement of both systems at one sample by just moving the spot of the exciting laser. Wave–vector transfers can be realized by tilting the sample normal with respect to the directions of incoming and scattered light. In the case of wires, the wave-vector q can be transferred, e.g., in y direction, perpendicular to the wires, or in x direction, parallel to the wires (cf. Chap. 6). A semi-transparent Titanium gate covers the sample surface in the wire area as well as in the double–quantum–well area. By applying an electrical field between the alloyed contacts of the 2DES and the gate, the electron density and the symmetry of the double–layered systems in z direction can be controlled at the same time. Due to the asymmetry of the double quantum well potential (see Fig. 7.2b), the wave function of the lower level is located dominantly in the left and the wave function of the upper level in the right well. The electron density of the 2DES can be varied between 5 and 9×10^{11} cm^{-2}. In the high density range, the two lowest tunnelling-split subbands of the double–layered potential are occupied. By decreasing the density, the occupation changes, which can be deduced from the energy variation of the 2D ISP, which originates from intersubband transitions.

The period and geometrical wire width of the investigated sample are 700 nm and 250 nm, respectively. Due to lateral depletion, the *electronic* wire width is smaller. Comparing the wave–vector dependent measurements of confined OP modes (not shown here) with the hydrodynamical model of Eliasson et al. [27], the effective electronic wire width, a, can be deduced. For an applied gate voltage of -400 mV ($+700$ mV) we get $a = 130$ nm ($a = 150$ nm) and the one-dimensional density $N_{1D} = 2.8 \times 10^6$ cm^{-1} ($N_{1D} = 5.8 \times 10^6$ cm^{-1}) with $N_{1D} \simeq a \cdot N_{2D}$.

Figure 7.10 shows a series of measured Raman spectra of electronic excitations. The left (right) figure shows spectra in polarized (depolarized) configuration, i.e., the polarization of the incoming and scattered light are parallel (perpendicular) to each other. The external electric field is varied from

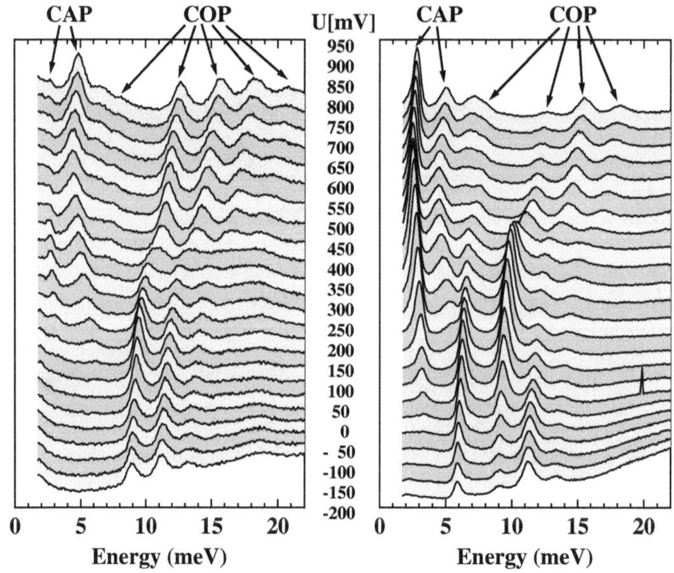

Fig. 7.10. Polarized (*left*) and depolarized (*right*) spectra of electronic excitations in a tunnelling–coupled GaAs–AlGaAs quantum wire array. The spectra are shifted against each other for clarity

950 mV down to −200 mV in steps of 50 mV. In the experiments presented here, a wave vector $q = 0.81 \times 10^5$ cm^{-1} was transferred perpendicular to the wire direction (y direction) to effectively induce electron motion in the direction of the lateral confinement. The energy of the incoming laser light was 1606 meV. It clearly can be seen that there are two different types of excitations. On the one hand there are excitations which are shifting to lower energies with decreasing gate voltage (labelled COP); on the other hand there are excitations which are shifting first to *lower* and then to *higher* energies when the gate voltage is decreased (labelled CAP). We infer that the CAP are confined acoustic, and the COP are confined optical plasmons of the tunnelling–coupled quantum wires (see discussion further below). Macroscopically, for our experimental geometry, the carriers are oscillating in both layers perpendicular to the wire direction. In the case of COP *in phase* in both layers, and *out of phase* in the case of CAP. Measurements on the wire sample, which were performed in plane wave geometry, i.e., transferring the wave vector parallel to the wires (x direction) support this interpretation: We observe modes, which can be identified as acoustic intrasubband plasmons and optical intrasubband plasmons by their wave vector dispersions, which are linear in q and proportional to $\sqrt{q}$, respectively.

Up to this point it is not clear, why the observed excitations in Fig. 7.10 do not show distinct polarization selection rules, as, e.g., reported for single–layered quantum wires [28]. There, charge–density excitations should only be observable in polarized configuration if the wave vector was transferred perpendicular to the wire direction. We think that in our case, both, resonant scattering, and near–field effects within the wire array play an important role and need to be further investigated theoretically.

In Fig. 7.11, the measured excitation energies vs. gate voltage are depicted. Black squares represent excitations which are detected in polarized, grey circles in depolarized configuration. The indices give the change, Δj, in the one–dimensional subband quantum number, j, for the transitions, which contribute predominantly to the observed excitations. The insets show a simple scheme of the macroscopic density distributions for the first localized optical (COP_1, upper scheme) and acoustic (CAP_1, lower scheme) Q1D intersubband plasmons in a double–layered quantum wire. The arrows indicate

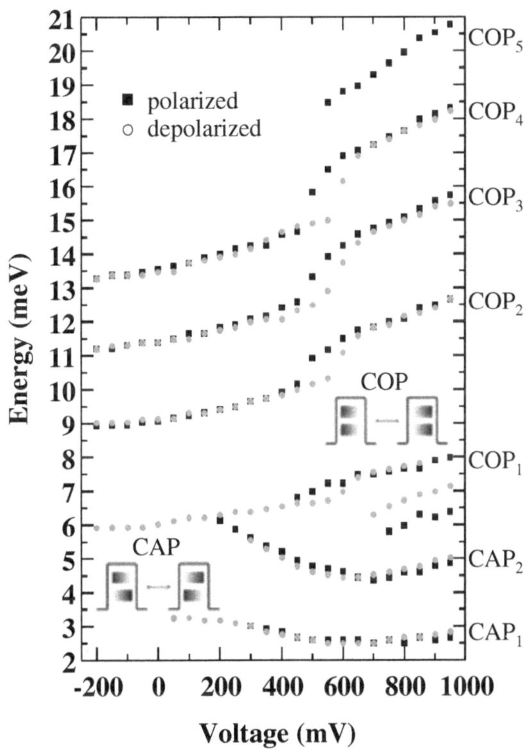

Fig. 7.11. Measured peak positions of the confined optical ($COP_{\Delta j}$) and acoustic ($CAP_{\Delta j}$) intersubband plasmons in a double–layered 1DES vs. gate voltage. The insets show a scheme of the macroscopic density distribution for the first localized optical (*upper*) and acoustic (*lower*) intersubband plasmons

the two extreme situations, between which the system oscillates macroscopically. Calculations of density distributions in weakly–coupled, double–layered 1DES had been performed by Steinebach et al. [29].

The striking feature observed here is the different behavior of optical modes in comparison to the acoustic modes. We will first concentrate on the characteristics of the excitation energies for both excitation types. Those can be explained in a simple macroscopic picture as follows: In the case of $\mathrm{COP}_{\Delta j}$, the carriers in both 1D layers oscillate in phase perpendicular to the wire direction. The spatial separation of the electrons in both layers is essentially given by the tunnelling barrier with a width of 1 nm. The higher the charge density of the wires, the larger is the Coulomb interaction. Therefore, the energies of the $\mathrm{COP}_{\Delta j}$ increase with increasing charge density, i.e., with increasing gate voltage. The excitation energies are dominated by the total carrier density and are nearly insensitive to the variation of the double–well potential from symmetric to asymmetric and vice versa. The energies of the $\mathrm{CAP}_{\Delta j}$ are lower in comparison to the optical excitations for the whole voltage range. In the extreme situations, sketched in the inset of Fig. 7.11, due to the out of phase oscillation, one part of the carrier density is located in one well on one side of the wire, while the other is located in the other well just on the opposite side of the wire. Therefore, generally, the Coulomb interaction is weaker for the acoustic modes than for the optical modes and hence the excitation energies are smaller. This effect approaches a maximum when the double quantum well potential is symmetric. Then, the effective charge displacement is just zero, since half of the density is moving to one side while the other half is moving in the opposite direction. In this case, we expect a minimum in the excitation energy, since the effect of the Coulomb interaction is minimal. This is observed experimentally in the excitation energies of the acoustic modes in Fig. 7.11. In this simple picture, however, one can not explain why the minima for the CAP_1 and CAP_2 do not occur at exactly the same voltage.

In the case of the $\mathrm{COP}_{\Delta j}$, furthermore, discontinuities are clearly observed in both polarization configurations at gate voltages between 450 and 650 mV. It seems that the strengths of these discontinuities increase with increasing excitation energies, and they might be also present for the $\mathrm{CAP}_{\Delta j}$ but are not clearly resolvable, since their excitation energies are smaller. So far, there is no definitive explanation of this effect. It might be due to a self–consistent carrier–density redistribution when the second 2D subband, i.e., the subband with asymmetric wave function, starts to be filled with electrons. There is, however, no convincing proof of this situation.

In conclusion, in this section first experimental observations of confined optical and acoustic intersubband plasmons in double–layered quantum wires in the regime of strong tunnelling coupling were reported. By applying an external electric field, which influences on the one hand the 1D electron density and on the other hand the symmetry of the double–quantum–well structure, we could differentiate between the two excitation types by the variation of their energies with gate voltage.

References

1. S. Das Sarma and A. Madhukar: Phys. Rev. B **23**, 805 (1981)
2. G. Fasol, N. Mestres, H. P. Hughes, A. Fischer, and K. Ploog: Phys. Rev. Lett. **56**, 2517 (1986)
3. D. S. Kainth, D. Richards, H. P. Hughes, M. Y. Simmons, and D. A. Ritchie: Phys. Rev. B **57**, R2065 (1998); D. S. Kainth, D. Richards, A. S. Bhatti, H. P. Hughes, M. Y. Simmons, E. H. Linfield, and D. A. Ritchie: Phys. Rev. B **59**, 2095 (1999)
4. G. S. Boebinger, H. W. Jiang, L. N. Pfeiffer, and K. W. West: Phys. Rev. Lett. **64**, 1793, (1990)
5. M. Hartung, A. Wixforth, K. L. Campman, and A. C. Gossard: Solid State Electronics **40**, 113 (1996)
6. R. Decca, A. Pinczuk, S. Das Sarma, B. S. Dennis, L. N. Pfeiffer, and K. W. West: Phys. Rev. Lett. **72**, 1506 (1994)
7. V. Pellegrini, A. Pinczuk, B. S. Dennis, A. S. Plaut, L. N. Pfeiffer, and K. W. West: Phys. Rev. Lett. **78**, 310 (1997)
8. A. Sawada, Z. F. Ezawa, E. Ohno, Y. Horikoshi, Y. Ohno, S. Kishimoto, F. Matsukura, M. Yasumoto, and A. Urayama: Phys. Rev. Lett. **80**, 4534 (1998)
9. I. B. Spielman, J. P. Eisenstein, L. N. Pfeiffer, and K. W. West: Phys. Rev. Lett. **84**, 5808 (2000)
10. S. Holland, C.-M. Hu, Ch. Heyn, and D. Heitmann: Phys. Rev. B **66**, 073305 (2002)
11. M.-T. Bootsmann, C.-M. Hu, Ch. Heyn, D. Heitmann, and C. Schüller: Phys. Rev. B **67**, 121309(R) (2003)
12. T. Chakraborty and P. Pietilainen: Phys. Rev. Lett. **59**, 2784 (1987)
13. A. H. MacDonald, P. M. Platzman, and G. S. Boebinger, Phys. Rev. Lett. **65**, 775 (1990)
14. X. G. Wen and A. Zee, Phys. Rev. Lett. **69**, 1811 (1992)
15. K. Yang, K. Moon, L. Zheng, A. H. MacDonald, S. M. Girvin, D. Yoshioka, and S.-C. Zhang: Phys. Rev. Lett. **72**, 732 (1994)
16. A. Stern, S. Das Sarma, M. P. A. Fisher, and S. M. Girvin: Phys. Rev. Lett. **84**, 139 (2000)
17. S. Das Sarma and E. H. Hwang: Phys. Rev. Lett. **81**, 4216 (1998)
18. P. G. Bolcatto and C. R. Proetto: Phys. Rev. Lett. **85**, 1734 (2000)
19. C.-M. Hu and D. Heitmann: Appl. Phys. Lett. **77**, 1475 (2000)
20. C.-M. Hu, C. Schüller, and D. Heitmann: Phys. Rev. B **64**, 073303 (2001)
21. Maik–Thomas Bootsmann: *Elektronische Raman–Spektroskopie an Quantenpunkten, Doppelquantenfilmen und Doppelquantenpunkten* (Shaker, Aachen, 2003)
22. T. Demel, D. Heitmann, P. Grambow, and K. Ploog: Phys. Rev. B **38**, 12732 (1988)
23. V. Shikin, T. Demel, and D. Heitmann: Phys. Rev. B **46**, 3971 (1992)
24. Q. P. Li, and S. Das Sarma: Phys. Rev. B **43**, 11768 (1991)

25. C. Steinebach, D. Heitmann, and V. Gudmundsson: Phys. Rev. B **58**, 13944 (1998)
26. F. Stern: Phys. Rev. Lett. **18**, 546 (1967)
27. G. Eliasson, J.-W. Wu, P. Hawrylak, and J. J. Quinn: Solid State Commun. **60**, 41 (1986)
28. C. Dahl, B. Jusserand, and B. Etienne: Phys. Rev. B **51**, 17211 (1995)
29. C. Steinebach, D. Heitmann, and V. Gudmundsson,: Phys. Rev. B **56**, 6742 (1997)

8 Inelastic Light Scattering in Microcavities

8.1 Introduction

A semiconductor microcavity is an optical resonator, where the mirrors consist of alternating layers of two different semiconductors with different refractive indices, e.g., GaAs and AlAs, which have thicknesses of a quarter wavelength each. The resonator itself is called *spacer*, and has a thickness of a few half wavelengths. The electric fields of the light waves, and hence the light–matter interaction can be modified significantly inside a microcavity. In the past decades, a number of sophisticated experiments have been reported that took advantage of the strongly enhanced electric field inside the spacer of a planar semiconductor microcavity. A prominent example is the enhanced exciton–photon coupling, resulting in an enlarged Rabi splitting, in planar microcavities containing *undoped* quantum wells [1]. Subsequently, a wealth of theoretical and experimental work on exciton polaritons in semiconductor microcavities, e.g., about the influence of a magnetic field [2], or coupling between different microcavities [3], followed.

Fainstein et al. [4] impressively demonstrated for the first time that the enhanced electric field inside a microcavity can also enhance optical–phonon Raman scattering by over 4 orders of magnitude. Even stronger enhancements could be reached by using the cavity polariton mode as a resonant intermediate state in the scattering process [5]. In all these investigations, excitonic or polaritonic effects caused by the interaction of photo–excited carriers in *undoped* structures with the photonic cavity mode were studied.

In this chapter, we will discuss the investigation of *electronic excitations* of a 2DES, embedded inside a planar semiconductor microcavity [6, 7]. It will be shown that the inelastic light scattering by electronic intersubband excitations can be enhanced by about three orders of magnitude under conditions of an optical double resonance [4], where the incoming photons as well as the inelastically scattered photons are in resonance with the cavity mode, compared to the single–resonance case, where only the laser photons are in resonance with the cavity mode. Moreover, since a high–quality cavity with a cavity–mode width smaller than the widths of the electronic excitations was used, it is possible to selectively enhance distinct parts of the excitations. These investigations may offer the opportunity to use the presented method in the future as some kind of a selective spectrometer to enhance

Christian Schüller: *Inelastic Light Scattering of Semiconductor Nanostructures*
STMP **219**, 161–168 (2006)
DOI 10.1007/3-540-36526-5_8

electronic elementary excitations. This would enable one to study, e.g., excitations in quantum–wire or quantum–dot structures containing only few electrons, which might be too weak for a direct observation.

8.2 2DES Inside a Semiconductor Microcavity

We will start by introducing the sample structure used by T. Kipp et al. [6, 7]. The MBE grown microcavity samples consist of a $Al_{0.2}Ga_{0.8}As$ λ cavity (spacer) with AlAs-$Al_{0.4}Ga_{0.6}As$ quarter–wave layers on both sides (24 on the top, and 33 on the bottom side), which act as distributed Bragg reflectors (DBR's). The finesse of the microcavities is between about 1200 and 2400. In the center of the spacer, there is a one–sided modulation–doped GaAs quantum well with 30 nm well width grown. A Si–doped layer is separated from the well by a 20 nm spacer. The sample structure is schematically shown in Fig. 8.1a. Figure 8.1b shows the calculated distribution of the electric field amplitudes inside the cavity. To allow for simultaneous transmission and Raman experiments, the GaAs substrate was completely removed from the back side of the sample. After the removal of the substrate, the only GaAs which was left in the sample was the 30 nm quantum well and an undoped 2 nm cap layer. By magneto–luminescence measurements, the carrier density inside the well could be determined to be 5.3×10^{11} cm^{-2}. The Raman experiments were performed using a tunable Ti:sapphire laser and a triple Raman spectrometer with liquid–nitrogen cooled CCD camera. The sample

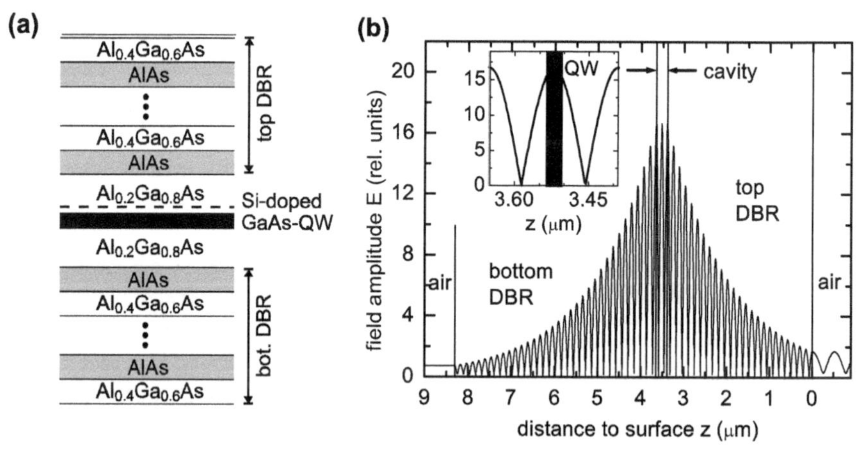

Fig. 8.1. (**a**) Layer sequence of a sample containing a modulation–doped single quantum well embedded inside a semiconductor microcavity. (**b**) Calculated electric–field amplitudes inside the microcavity sample. The inset shows a magnification of the field distribution inside the spacer. The position of the quantum well is marked by the vertical bar

was immersed in superfluid liquid helium at a temperature of about 1.6 K. The scattering geometry is shown in Fig 8.3a. $\mathbf{k}_i$ and $\mathbf{k}_s$ label the wave vectors of the incident and scattered beams, respectively. In the quasi–backscattering geometry there was generally a finite angle between the sample normal and the direction of incident (θ_i) and scattered light (θ_s), as well as between both beams ($\theta_i - \theta_s \neq 0$). To most accurately fix the direction of the scattered light, which was collected by a lens, a 2 mm aperture at a distance of about 0.5 m from the sample was used (see Fig. 8.3a). The thickness of the MBE layers and thus the cavity–mode wavelength decreases from the center of the wafer towards the edges. So, by laterally moving the laser spot across the sample, the cavity mode could be tuned. For the Raman experiments, a cavity–mode energy below the GaAs bandgap was choosen to separate the cavity-related resonance from electronic resonances of the quantum well.

8.3 Optical Double–Resonance Experiments

First, we want to elucidate the so–called single–resonance case. Figure 8.2a shows spectra of intersubband excitations of the 2DES for exact backscattering geometry, i.e., the directions of incident and scattered light were parallel and antiparallel to the sample normal, respectively. We note that these spectra were taken without the aperture to get higher intensities. For parallel polarizations of incident and scattered light (polarized configuration) the in-

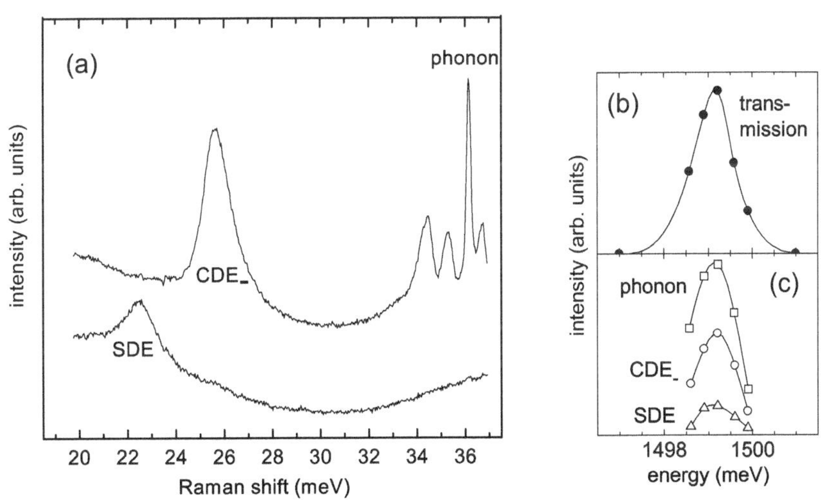

Fig. 8.2. (**a**) Polarized (*upper spectrum*) and depolarized (*lower spectrum*) spectra of intersubband excitations in a 30 nm quantum well for single–resonance conditions. (**b**) Transmitted intensity, and, (**c**) intensities of the observed excitations versus laser energy

tersubband charge-density excitation (CDE) is observed [8]. This mode is affected by direct and exchange Coulomb interactions and is therefore shifted to higher energies compared to the single–particle subband spacing of the system in its ground state. The intersubband CDE is split into two modes, CDE_- and CDE_+, due to the coupling with the LO phonon of the polar lattice (cf. Sect. 3.3.3). The uncoupled LO phonon which is also observed in the polarized spectrum, originates from the undoped GaAs buffer layer before complete removal. In the depolarized configuration, i.e., crossed polarization directions, the spin–density excitation (SDE) is observed, which is shifted to lower energies since it is only affected by exchange interaction. These spectra were taken with the laser energy in resonance with the cavity mode at about 1499 meV. In Fig. 8.2c, the intensities of the observed excitations – SDE and CDE_- – for this single–resonance case are shown versus laser energy. Simultaneously to the Raman experiments, the intensity of the light which was transmitted through the sample was measured (Fig. 8.2b). This clearly shows that the resonance maxima of the excitations are directly related to the cavity–mode energy. An intrinsic *electronic* resonance of the quantum well can be excluded. The first electronic resonance via a quantum–well exciton as an intermediate state in the scattering process occurs at the transition energy from the highest hole subband to the first excited electron subband, at much higher energies, which was measured at about 1525 meV.

In the double–resonance experiment, both the laser energy *and* the energy of the scattered light are in resonance with the cavity mode. We will describe further below how this situation can be accomplished. Figure 8.3b

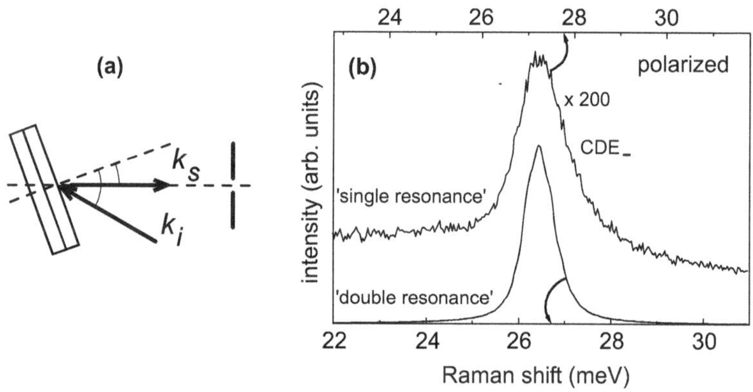

Fig. 8.3. (**a**) Schematic picture of the scattering geometry for double–resonance experiments. (**b**) Polarized Raman spectra for single–resonance and double–resonance configurations. The spectra were taken for incident angles $\theta_i = 30.3°$ and $\theta_i = 37.3°$ for the single–resonance and double–resonance case, respectively. The scattering angles were $\theta_s = \theta_i - 26.8°$. For the single-resonance spectrum a background was substracted before multiplication

shows the striking result of this experiment. In Fig. 8.3b, polarized spectra of the intersubband CDE_- are compared for the single-resonance and the double-resonance configuration. Now, the scattered light is collected through a 2 mm aperture, as shown in Fig. 8.3a, to accurately fix the direction with respect to the direction of the incoming laser beam. There are two remarkable observations which we will discuss in the following. First, the intensity in double resonance is about 3 orders of magnitude larger than in single resonance. Note that the base line is the same for both spectra. The single–resonance spectrum is multiplied by a factor of 1500 so that the maximum intensities in both spectra, single–resonance and double–resonance, are the same. We will come back to the electronic background, which is visible in the single–resonant spectrum, below when we describe the double–resonance case. In the following we will discuss some electrodynamic arguments considering this drastic effect. To estimate the enhancement of the electric field inside the cavity, we have performed calculations of the field distribution inside the cavity by solving Maxwell's equations for the multi-layer system. Figure 8.1b shows the results of this calculation. It can be seen that the field amplitude is maximal in the center of the λ cavity and decays exponentially inside the DBR's. The calculations were performed assuming s–polarized incident light and a tilt angle of 37.3°. These are the conditions where double resonance is achieved in the experiments (see below). From these calculations we find that the electric field is amplified by a factor of about 16 in the center of the cavity, where the quantum well containing the 2DES is located. In a simple approach, we want to consider first these electric field amplitudes inside the cavity. The scattering amplitude, $\gamma_{\alpha\beta}$, for inelastic light scattering by electronic excitations is given by [9] (cf. Sect. 4.2.2)

$$\gamma_{\alpha\beta} \sim \sum_{\beta'} \frac{\langle\beta|\mathbf{e}_s E_s|\beta'\rangle\langle\beta'|\mathbf{e}_i E_i|\alpha\rangle}{\hbar\omega_i - \epsilon_\beta + \epsilon_{\beta'}} , \tag{8.1}$$

where $\mathbf{e}_i$ ($\mathbf{e}_s$) and E_i (E_s) mean the polarization and electric–field amplitude of the incident (scattered) light, respectively. $|\alpha\rangle$, $|\beta\rangle$, and $|\beta'\rangle$ are single–particle states of the quantum well, and ϵ_i are the corresponding energies. $\hbar\omega_i$ is the energy of the incident laser photons. The resonance denominator in (8.1) plays a relevant role for *electronic* resonances via real transitions between valence– and conduction–band states of the quantum well, which shall not be discussed here. For the purely *optical* resonances considered here, mediated by the electro–dynamic enhancement of the field amplitude inside the optical resonator, the numerator is important since here the electric–field amplitudes E_i and E_s of the incident and inelastically scattered radiations enter. From this formula we can in this simple approach estimate that if in the double–resonance case, besides E_i, also E_s is enhanced by about one order of magnitude, then the scattered intensity, which is proportional to $|\gamma_{\alpha\beta}|^2$, is enhanced by about 2 orders of magnitude, compared to the single–resonance case. However, this consideration would describe only the effects inside the

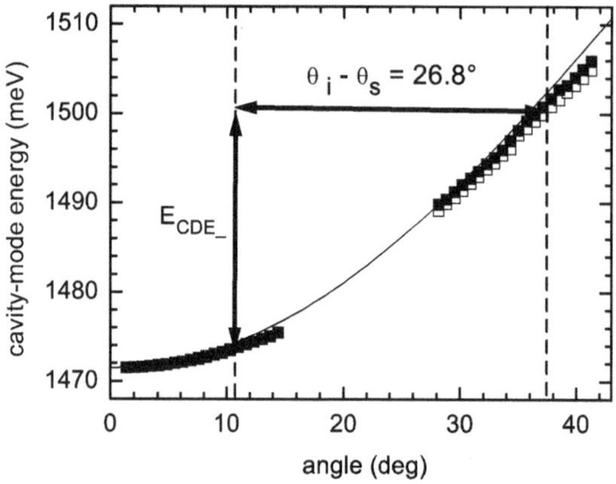

Fig. 8.4. Calculated cavity-mode energy versus incident angle (*solid line*). The symbols are the experimentally determined mode positions for p-polarized (*solid symbols*) and s-polarized (*open symbols*) light

cavity sufficiently and not the intensity of the signal which is measured outside the cavity . For a comparison of the double–resonance to a non–resonance case without cavity, this approach might be useful [4] since in that case both the incoming and scattered radiation are in resonance with a cavity mode. For the comparison of the single–resonance and the double–resonance case in backscattering geometry, however, one has to take into account that in single resonance the scattered light is out of resonance with the vertical cavity and therefore should exhibit strong extinction. For a more appropriate description one has to consider for that case an emitting dipole at the location of the quantum well inside the cavity. With this, one has to compute the radiation pattern of the dipole inside the cavity in order to determine the intensity emitted out of the cavity in the direction where the scattered light is collected [10, 11]. We assume that in our case, because of the extinction of the scattered light in single resonance, the experimentally detected enhancement is about one order of magnitude larger than the one we got from the simple approach used above.

The second remarkable observation in Fig. 8.3b is that the width of the excitation in double resonance is about a factor of 2 smaller than in the single–resonance case. To elaborate on this in more detail, we will describe now how we accomplish double resonance in modification of the experiment of Fainstein et al. on optical phonons [4]. Figure 8.4 displays calculated (solid line) and experimental (symbols) cavity–mode energies versus incident angle, where the angle is measured between the sample normal and the direction of the light. The experimental mode positions do not exactly match the calculated curve. We assume that in the experiment the laser spot was not located exactly on the rotation axis of the sample so that by tilting the sample also

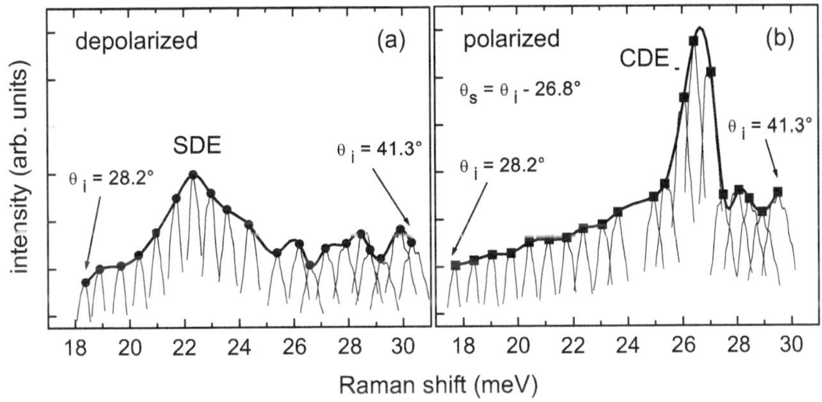

Fig. 8.5. (**a**) Depolarized and (**b**) Polarized double–resonant Raman spectra in the range of the electronic intersubband excitations. For clarity, only the spectral ranges around the maxima are shown. The incident angles were varied as displayed in the figures. The laser energies varied between 1489.84 meV and 1505.91 meV for the depolarized spectra, and between 1489.16 meV and 1504.95 meV for the polarized spectra. The maxima have been connected by an envelope curve

the position of the laser spot changed slightly. We estimate that a variation of the position on the sample surface of about 0.6 mm could account for the difference. Note that also the difference between p– (solid symbols in Fig. 8.4) and s–polarized light (open squares in Fig. 8.4) in the experimental cavity–mode energies is much larger (about 1 meV) than the calculated difference (much less than 1 meV), which is not resolvable in Fig. 8.4. We assume that also this difference between experiment and theory is due to a slight displacement of the laser spot on the sample by switching between the two polarizations. We started the double–resonance experiment by having the scattered light parallel to the sample normal and the incident light at an angle of 26.8°. After that, the sample was tilted as shown in Fig. 8.3a in small steps of 0.7°, while the angle between incident and scattered light was kept constant at 26.8°. For each tilt angle, the laser energy was tuned in resonance with the cavity mode. This was monitored by determining the maximum of the intensity which was transmitted through the sample. Double resonance is reached when the difference in the cavity–mode energies of incident and scattered light just matches the energy of an excitation. This situation is indicated by the two vertical dotted lines in Fig. 8.4, which correspond to a tilt angle of 10.7°. Figure 8.5 displays the experimental results of such a measurement. It shows series of depolarized (Fig. 8.5a) and polarized (Fig. 8.5b) inelastic light scattering spectra in the range of tilt angles θ_s where double resonance is achieved. For clarity, only the spectral ranges around the maxima in the Raman spectra are shown as thin solid lines. The spectra were recorded for tilt angles between 1.4° and 14.5° in steps of about 0.7°.

Note that each spectrum belongs to a different laser energy. The solid symbols mark the maxima of the Raman spectra and the thick solid lines which connect the symbols serve as a guide to the eye to represent the 'envelopes' of the measured spectra. By comparing these envelope curves to the spectra in Fig. 8.2a, measured in single resonance, one can see that they nearly reproduce the line shapes. It can be seen that also the electronic background, which we mentioned before in the single–resonance spectrum in Fig. 8.2a, is enhanced in the double–resonance experiment. Thus, by a distinct choice of the angle arrangement of incident and scattered beams, we are able to selectively enhance parts of the, in this case, inhomogeneously broadened excitations. The spectral width of the enhanced part is determined by the spectral width of the cavity mode.

In conclusion, in this chapter, an optical resonance of electronic excitations of a modulation–doped quantum well embedded inside a microcavity was presented. By a variation of incidence and scattering angles, double–resonance conditions could be achieved in a spectral range, which is determined by the width of the cavity mode. This enables one to selectively enhance parts of electronic excitations by about 3 orders of magnitude compared to a single resonance.

References

1. C. Weisbuch, M. Nishioka, A. Ishikawa, and Y. Arakawa: Phys. Rev. Lett. **69**, 3314 (1992)
2. J. Tignon, P. Voisin, C. Delalande, M. Voos, R. Houdre, U. Oesterle, and R. P. Stanley: Phys. Rev. Lett. **74**, 3967 (1995)
3. R. P. Stanley, R. Houdre, U. Oesterle, M. Gailhanou, and M. Ilegems: Appl. Phys. Lett. **65**, 2093 (1994)
4. A. Fainstein. B. Jusserand, and V. Thierry-Mieg: Phys. Rev. Lett. **75**, 3764 (1995)
5. A. Fainstein. B. Jusserand, and V. Thierry-Mieg: Phys. Rev. Lett. **78**, 1576 (1997)
6. T. Kipp, L. Rolf, C. Schüller, D. Endler, Ch. Heyn, and D. Heitmann: Phys. Rev. B **63**, 195304 (2001)
7. T. Kipp, L. Rolf, C. Schüller, D. Endler, Ch. Heyn, and D. Heitmann: Physica E **13**, 408 (2002)
8. For an overview see: A. Pinczuk and G. Abstreiter in: *Light Scattering in Solids V*, Topics in Applied Physics Vol. 66, eds. M. Cardona and G. Güntherodt (Springer, Berlin, 1988) p. 153
9. F. A. Blum: Phys. Rev. B **1**, 1125 (1970)
10. A. Fainstein, B. Jusserand, and V. Thierry-Mieg: Phys. Rev. B **53**, R13287 (1996)
11. A. Fainstein and B. Jusserand: Phys. Rev. B **57**, 2402 (1998)

Part III

Appendix

A Kronecker Products of Dipole Matrix Elements I

In this chapter, Kronecker products corresponding to dipole matrix elements of Bloch functions of the Γ_6 conduction–band edge and the Γ_7 split–off valence–band edge are calculated. The corresponding band–edge Bloch functions are listed in Table 4.29 on page 71. The calculation shall be explained by the following example:

$$\langle S \uparrow |\boldsymbol{p}|V2\rangle\langle V2|\boldsymbol{p}|S \uparrow\rangle \; . \tag{A.1}$$

The momentum matrix element $\langle S \uparrow |\boldsymbol{p}|V2\rangle$ is given by

$$\begin{aligned}\langle S \uparrow |\boldsymbol{p}|V2\rangle &= \frac{i}{\sqrt{3}}\langle S \uparrow |\boldsymbol{p}| - (X - iY) \uparrow + Z \downarrow\rangle = \\ &= \frac{1}{\sqrt{3}}\begin{pmatrix} -i\langle S|p_x|X\rangle \\ -\langle S|p_y|Y\rangle \\ 0 \end{pmatrix} = \frac{1}{\sqrt{3}}\begin{pmatrix} P \\ -iP \\ 0 \end{pmatrix} ,\end{aligned} \tag{A.2}$$

where P is defined as

$$P := -i\langle S|p_x|X\rangle = -i\langle S|p_y|Y\rangle = -i\langle S|p_z|Z\rangle \; . \tag{A.3}$$

Furthermore, we have

$$\langle u_i|\boldsymbol{p}|u_k\rangle = \langle u_k|\boldsymbol{p}|u_i\rangle^* \; . \tag{A.4}$$

With (A.2) and (A.4) we get for the product (A.1):

$$\langle S \uparrow |\boldsymbol{p}|V2\rangle\langle V2|\boldsymbol{p}|S \uparrow\rangle = \frac{P^2}{3}\begin{pmatrix} 1 \\ -i \\ 0 \end{pmatrix}(1, i, 0) = \frac{P^2}{3}\begin{pmatrix} 1 & i & 0 \\ -i & 1 & 0 \\ 0 & 0 & 0 \end{pmatrix} . \tag{A.5}$$

Analogously, we receive at the remaining Kronecker products, which are listed in Table B1.

Table B1. Kronecker products of dipole matrix elements between the Γ_6 and Γ_7 band–edge Bloch functions

$\otimes$	$\langle V1\vert\boldsymbol{p}\vert S\uparrow\rangle$	$\langle V1\vert\boldsymbol{p}\vert S\downarrow\rangle$
$\langle S\uparrow\vert\boldsymbol{p}\vert V1\rangle$	$\frac{P^2}{3}\begin{pmatrix}0&0&0\\0&0&0\\0&0&1\end{pmatrix}$	$\frac{P^2}{3}\begin{pmatrix}0&0&0\\0&0&0\\1&-i&0\end{pmatrix}$
$\langle S\downarrow\vert\boldsymbol{p}\vert V1\rangle$	$\frac{P^2}{3}\begin{pmatrix}0&0&1\\0&0&i\\0&0&0\end{pmatrix}$	$\frac{P^2}{3}\begin{pmatrix}1&-i&0\\i&1&0\\0&0&0\end{pmatrix}$

$\otimes$	$\langle V2\vert\boldsymbol{p}\vert S\uparrow\rangle$	$\langle V2\vert\boldsymbol{p}\vert S\downarrow\rangle$
$\langle S\uparrow\vert\boldsymbol{p}\vert V2\rangle$	$\frac{P^2}{3}\begin{pmatrix}1&i&0\\-i&1&0\\0&0&0\end{pmatrix}$	$\frac{P^2}{3}\begin{pmatrix}0&0&-1\\0&0&i\\0&0&0\end{pmatrix}$
$\langle S\downarrow\vert\boldsymbol{p}\vert V2\rangle$	$\frac{P^2}{3}\begin{pmatrix}0&0&0\\0&0&0\\-1&-i&0\end{pmatrix}$	$\frac{P^2}{3}\begin{pmatrix}0&0&0\\0&0&0\\0&0&1\end{pmatrix}$

B Kronecker Products of Dipole Matrix Elements II

At the fundamental bandgap E_0 we have for Q2D systems generally a mixing of heavy and light–hole states. In this case, the dipole matrix elements between a valence–band state $\psi_{h_i,\boldsymbol{k}_{||}}(\boldsymbol{r})$ and a conduction–band state $\psi_{c_j,\boldsymbol{k}_{||}}(\boldsymbol{r})$ is given by

$$\begin{aligned}
\langle h_i, \boldsymbol{k}_{||}|\boldsymbol{p}|c_j, \boldsymbol{k}_{||}\rangle = & \int \chi^*_{3/2,h_i,\boldsymbol{k}_{||}}(z)\, \chi_{1/2,c_j}(z)\, dz \left\langle \frac{3}{2}, \ \frac{3}{2}|\boldsymbol{p}|\frac{1}{2}, \ \frac{1}{2}\right\rangle + \\
& + \int \chi^*_{1/2,h_i,\boldsymbol{k}_{||}}(z)\, \chi_{1/2,c_j}(z)\, dz \left\langle \frac{3}{2}, \ \frac{1}{2}|\boldsymbol{p}|\frac{1}{2}, \ \frac{1}{2}\right\rangle + \\
& + \int \chi^*_{-1/2,h_i,\boldsymbol{k}_{||}}(z)\, \chi_{1/2,c_j}(z)\, dz \left\langle \frac{3}{2}, -\frac{1}{2}|\boldsymbol{p}|\frac{1}{2}, \ \frac{1}{2}\right\rangle + \\
& + \int \chi^*_{-3/2,h_i,\boldsymbol{k}_{||}}(z)\, \chi_{1/2,c_j}(z)\, dz \left\langle \frac{3}{2}, -\frac{3}{2}|\boldsymbol{p}|\frac{1}{2}, \ \frac{1}{2}\right\rangle , \qquad \text{(B.1)}
\end{aligned}$$

where a Bloch state $|J, J_z\rangle$ is characterized by its total angular momentum quantum number J and the corresponding z component J_z. The $\chi_{3/2,h_i,\boldsymbol{k}_{||}}(z)$ and $\chi_{1/2,c_j}(z)$ are the envelope functions of the corresponding hole and electron wave functions, respectively. Following the envelope–function approximation, we assume that the dipole operator $\boldsymbol{p}$ operates on the lattice–periodic Bloch functions $|J, J_z\rangle$, only. The relevant dipole matrix elements $\langle J, J_z|\boldsymbol{p}|J', J'_z\rangle$ between Γ_8 and Γ_6 band–edge Bloch functions (see Table 1 on page 14) and their Kronecker products can be calculated as shown in appendix A. The resulting Kronecker products are listed in Tables C1 and C2.

Table C1. Kronecker products of dipole matrix elements between Γ_8 and Γ_6 band-edge Bloch functions

$\otimes$	$\langle \frac{1}{2}, -\frac{1}{2} \vert \boldsymbol{p} \vert \frac{3}{2}, -\frac{3}{2} \rangle$	$\langle \frac{1}{2}, -\frac{1}{2} \vert \boldsymbol{p} \vert \frac{3}{2}, -\frac{1}{2} \rangle$	$\langle \frac{1}{2}, -\frac{1}{2} \vert \boldsymbol{p} \vert \frac{3}{2}, \frac{1}{2} \rangle$	$\langle \frac{1}{2}, -\frac{1}{2} \vert \boldsymbol{p} \vert \frac{3}{2}, \frac{3}{2} \rangle$
$\langle \frac{3}{2}, -\frac{3}{2} \vert \boldsymbol{p} \vert \frac{1}{2}, \frac{1}{2} \rangle$	$\frac{P^2}{2} \begin{pmatrix} 1 & -i & 0 \\ i & 1 & 0 \\ 0 & 0 & 0 \end{pmatrix}$	$\frac{P^2}{\sqrt{3}} \begin{pmatrix} 0 & 0 & -i \\ 0 & 0 & 1 \\ 0 & 0 & 0 \end{pmatrix}$	$\frac{P^2}{2\sqrt{3}} \begin{pmatrix} 1 & i & 0 \\ i & -1 & 0 \\ 0 & 0 & 0 \end{pmatrix}$	0
$\langle \frac{3}{2}, -\frac{1}{2} \vert \boldsymbol{p} \vert \frac{1}{2}, -\frac{1}{2} \rangle$	$\frac{P^2}{\sqrt{3}} \begin{pmatrix} 0 & 0 & 0 \\ 0 & 0 & 0 \\ i & 1 & 0 \end{pmatrix}$	$\frac{2P^2}{3} \begin{pmatrix} 0 & 0 & 0 \\ 0 & 0 & 0 \\ 0 & 0 & 1 \end{pmatrix}$	$\frac{P^2}{3} \begin{pmatrix} 0 & 0 & 0 \\ 0 & 0 & 0 \\ i & -1 & 0 \end{pmatrix}$	0
$\langle \frac{3}{2}, \frac{1}{2} \vert \boldsymbol{p} \vert \frac{1}{2}, -\frac{1}{2} \rangle$	$\frac{P^2}{2\sqrt{3}} \begin{pmatrix} 1 & -i & 0 \\ -i & -1 & 0 \\ 0 & 0 & 0 \end{pmatrix}$	$\frac{P^2}{3} \begin{pmatrix} 0 & 0 & -i \\ 0 & 0 & -1 \\ 0 & 0 & 0 \end{pmatrix}$	$\frac{P^2}{6} \begin{pmatrix} 1 & i & 0 \\ -i & 1 & 0 \\ 0 & 0 & 0 \end{pmatrix}$	0
$\langle \frac{3}{2}, \frac{3}{2} \vert \boldsymbol{p} \vert \frac{1}{2}, -\frac{1}{2} \rangle$	0	0	0	0

Table C2. Kronecker products of dipole matrix elements between Γ_8 and Γ_6 band-edge Bloch functions

$\otimes$	$\langle \frac{1}{2}, \frac{1}{2} \vert \boldsymbol{p} \vert \frac{3}{2}, -\frac{3}{2} \rangle$	$\langle \frac{1}{2}, \frac{1}{2} \vert \boldsymbol{p} \vert \frac{3}{2}, -\frac{1}{2} \rangle$	$\langle \frac{1}{2}, \frac{1}{2} \vert \boldsymbol{p} \vert \frac{3}{2}, \frac{1}{2} \rangle$	$\langle \frac{1}{2}, \frac{1}{2} \vert \boldsymbol{p} \vert \frac{3}{2}, \frac{3}{2} \rangle$
$\langle \frac{3}{2}, -\frac{3}{2} \vert \boldsymbol{p} \vert \frac{1}{2}, \frac{1}{2} \rangle$	0	0	0	0
$\langle \frac{3}{2}, -\frac{1}{2} \vert \boldsymbol{p} \vert \frac{1}{2}, \frac{1}{2} \rangle$	0	$\frac{P^2}{6} \begin{pmatrix} 1 & -i & 0 \\ i & 1 & 0 \\ 0 & 0 & 0 \end{pmatrix}$	$\frac{P^2}{3} \begin{pmatrix} 0 & 0 & -i \\ 0 & 0 & 1 \\ 0 & 0 & 0 \end{pmatrix}$	$\frac{P^2}{2\sqrt{3}} \begin{pmatrix} 1 & i & 0 \\ i & -1 & 0 \\ 0 & 0 & 0 \end{pmatrix}$
$\langle \frac{3}{2}, \frac{1}{2} \vert \boldsymbol{p} \vert \frac{1}{2}, \frac{1}{2} \rangle$	0	$\frac{P^2}{3} \begin{pmatrix} 0 & 0 & 0 \\ 0 & 0 & 0 \\ i & 1 & 0 \end{pmatrix}$	$\frac{2P^2}{3} \begin{pmatrix} 0 & 0 & 0 \\ 0 & 0 & 0 \\ 0 & 0 & 1 \end{pmatrix}$	$\frac{P^2}{\sqrt{3}} \begin{pmatrix} 0 & 0 & 0 \\ 0 & 0 & 0 \\ i & -1 & 0 \end{pmatrix}$
$\langle \frac{3}{2}, \frac{3}{2} \vert \boldsymbol{p} \vert \frac{1}{2}, \frac{1}{2} \rangle$	0	$\frac{P^2}{2\sqrt{3}} \begin{pmatrix} 1 & -i & 0 \\ -i & -1 & 0 \\ 0 & 0 & 0 \end{pmatrix}$	$\frac{P^2}{\sqrt{3}} \begin{pmatrix} 0 & 0 & -i \\ 0 & 0 & -1 \\ 0 & 0 & 0 \end{pmatrix}$	$\frac{P^2}{2} \begin{pmatrix} 1 & i & 0 \\ -i & 1 & 0 \\ 0 & 0 & 0 \end{pmatrix}$

Index

acoustic intrasubband plasmons 155
acoustic plasmon 145
anticrossing 148
antistokes process 57

back contact 91, 112
band offsets 20
band structure 13
band structure calculation 14
band–edge Bloch functions 13–16, 22, 23, 71, 73, 74
bare potential 126
Ben Daniel–Duke boundary conditions 26
Ben Daniel–Duke model 24
Bernstein modes 130, 134, 138
Bloch ansatz 22
Bloch functions 22, 24, 71
Bloch states 10, 13, 75
Bloch waves 13
bonding p orbitals 11
bonding s orbitals 11
Bose–Einstein factor 61
boundary conditions 24, 25
Bravais lattice 10, 13
breathing mode 94, 97
Brillouin scattering 57
Brillouin zone 10, 12–15
bulk band structure 11, 12

cascade–like scattering process 80
charge–density excitations 41, 74, 127
charge–density fluctuations 70, 71, 73
charge–density waves 124, 127
chemical etching 35
confined acoustic plasmons 155
confined optical plasmons 155
confined plasmons 93, 124, 130–134
confinement potential 139
continuity conditions 24
correlations 26
Coulomb interaction 25, 26, 131, 138, 139
coupled bilayer system 153
coupled–layer plasmons 145
cyclotron resonance 80
cyclotron resonance frequency 101

deep etching 30
delta–doping layer 150
density of states 18
density pair operator 68, 70, 73
density–density correlation function 49, 61, 70, 131
density–fluctuation scattering process 89
density–functional theory 26
depletion length 122
depolarization field 45
depolarization shift 128, 131
depolarization–field effects 108
depolarized scattering geometry 74
diamond structure 10
dielectric function 138
dipole excitation 94, 95, 115
dipole matrix elements 71, 72, 75
dipole mode 97, 127
dipole moment 97
direct band gap 12
dissipation–fluctuation theorem 61
double quantum well 145, 146, 150, 152
double–layered 1DES 153
double–layered quantum wire 156
Dresselhaus effect 12
dry–etching plasma process 34

dynamical potential 51
dynamical structure factor 69

effective mass 10, 15, 24, 25
effective–mass approximation 14, 74
effusion cells 19
electromagnetic dipolar field 49
electron mobilities 22
electron–density fluctuations 68, 69, 73
electron–density oscillations 126
electron–phonon interaction 63
electronic band structure 9, 12
elementary electronic excitations 42
energy–density fluctuations 70, 74
envelope functions 22–24, 51, 75
envelope wave functions 28, 75, 76
envelope–function approximation 22
etch mask 30
etch–mask technique 32
etching processes 9
etching rate 33
exchange–correlation corrections 28
exchange–correlation effects 100
exchange–correlation energy 26
exchange–correlation potential 51
exclusion principle 88, 97
external potential 135

F. Stern 48
Fermi distribution function 50
Fermi liquid 138
Fermi wave vector 29, 42
fine–structure effects 96
frequency–dependent dielectric function 48, 53, 61, 138

G. Bastard 22
gate electrode 30
generalized correlation function 106
generalized Kohn's theorem 94, 115, 127, 134
generalized pair operator 74, 106
grating coupler 32, 78

Hartree approximation 25, 28
Hartree calculation 28, 131
Hartree energy 27
Hartree potential 26, 27
Hartree–Fock approximation 26, 130
heavy holes 12
heavy– and light–hole mixing 29
heterostructure growth 20
high–frequency dielectric constant 48, 53
Hohenberg 26
hydrodynamical model 124

III–V compound semiconductors 9, 11, 13
III–V semiconductors 9, 13, 20
independent–particle model 43
interference pattern 31
interferometric lithography 32
intersubband CDE 45
intersubband continuum 43
intersubband excitations 43
intersubband SDE 45
intraband plasmon 124, 139, 141
intrasubband continuum 42
intrasubband plasmon 138, 139
inversion symmetry 12
ion implantation 30
irreducible polarizability 138

J. M. Luttinger 14

k*p method 14
k*p theory 14
Kane model 14
Knudsen cells 19
Kohn 26
Kohn's mode 93, 97, 115, 124, 127, 133–135, 138
Kohn–Sham approximation 26
Kohn–Sham calculation 27, 28, 109
Kohn–Sham equation 26, 27
Kohn–Sham Hamiltonian 105
Kronecker products 71

L. D. Landau 47
Landau damping 47, 48, 127, 128
laser–interference lithography 31, 32, 34, 35, 139
lateral confinement 124, 131
lateral micro– and nanostructures 31
lateral nanostructures 9
lateral patterning 30

lateral quantum number 126, 131
lateral semiconductor nanostructures 30
lattice–periodic potential 63
layered heterostructures 9
lift–off process 30
light holes 12
lithography 9
LO phonon 54
local–density approximation 26
Luttinger liquid 138
Luttinger parameters 16
Luttinger–Kohn Hamiltonian 29

magnetoplasmons 131, 138
many–particle Schödinger equation 26
mean–field approaches 25
mesoscopic dots 89
metal–organic chemical–vapor deposition 18
modulation doping 21, 28
molecular–beam epitaxy 18
monopole excitation 94, 95
monopole mode 97

n–type doping 19
near–field effects 76
non–spinflip intersubband transitions 45
numerical diagonalization 25, 109, 115

one–electron Schrödinger equation 13
one–photon process 96
one–sided doping 22
optical intrasubband plasmons 155
optical phonon energies 53
optical plasmon 145
orbitals 11

p–type doping 19
parabolic approximation 24
parity selection rules 75, 76, 88, 95, 97, 112, 127
Pauli spin matrizes 13
Pauli spinmatrix 73
phonon–plasmon modes 53
photo–resist patterns 32
physical etching process 34
Pidgeon–Brown model 24
point group 9, 13
Poisson equation 22, 27
polar bulk semiconductors 53
polarizability function 138
polarization selection rules 44, 60, 71, 74, 76, 94, 102, 104, 126
polarized configuration 74
power spectrum 60
pseudo spin 145
pseudopotential calculation 14

Q0D system 17
Q1D system 17
Q2D system 17
quadrupole excitation 94, 95
quantum dots 18
quantum limit 138
quantum Monte Carlo calculations 26
quantum wells 18
quantum wires 18
quasi two–dimensional (Q2D) electron systems 9
quasiatomic orbitals 98, 101

radial quantum number 92
Raman scattering 57
Raman tensor 60
random–phase approximation 48
reactive–ion etching 35, 90
resist layer 30
resist mask 30
resist pattern 30, 33
RPA 48, 49, 131, 132, 138, 139, 147, 151

scattering amplitude 70, 71, 75, 76, 106
scattering cross section 66, 68–70, 89, 131
scattering mechanisms 62
Schrödinger equation 16
self–assembled growth 90
self–assembled quantum dots 36
self–consistent potential 28, 150
self–organized growth 9
semiconductor heterostructure 22, 25
semiconductor microcavity 161
semitransparent gate 147, 150

shadowing techniques 32
shallow etching 30
Sham 26
single–particle continua 42, 47, 123, 128
single–particle excitations 41
skipping–orbit mode 130, 137
skipping–orbit motion 137
Slater determinants 94, 111, 116
space group 9
spacer 22, 161
spacer layer 150
spectral differential cross section 60, 62
spin degeneracy 12, 15
spin dipole mode 97
spin monopole 94
spin monopole mode 97, 103
spin quadrupole 94
spin quadrupole mode 97
spin–charge separation 139
spin–density excitations 41, 74, 127
spin–density fluctuations 70, 71, 73, 74
spin–dipole excitation 94
spin–orbit coupling 11, 13, 30, 63
spin–orbit interaction 11, 13, 25, 74
spin–orbit splitting 13
split–off band 12
split–off valence band 15, 16, 24
Stokes process 57
Stranski–Krastanow growth 36
stripe pattern 33
structure factor 66
subband spacing 28
symmetry operations 13

TDLDA 48, 53, 139
TDLSDA 48
ternary alloy semiconductor 10
tight–binding methods 14
time–dependent local–density approximation 48
time–dependent local–spin–density approximation 48
Tomonaga–Luttinger model 139, 142
translational invariance 10, 23
tunneling barrier 150, 151
tunneling gap 146, 148
tunneling–coupled bilayer 145, 152
tunnelling-split subbands 154
two–component plasma 145
two–dimensional harmonic oscillator 92, 100
two–electron quantum dot 94
two–particle wavefunction 94
two–photon process 96
two–sided doping 22
two–subband occupation 147

under etching 34

vertical nanostructures 9, 22
virtual crystal approximation 10
von Neumann equation 49

W. Kohn 14
wave–vector dispersion 29
wet–chemical etching 34

Zincblende lattice 9
Zincblende structure 9

Springer Tracts in Modern Physics

177 **Applied Asymptotic Expansions in Momenta and Masses**
By Vladimir A. Smirnov 2002. 52 figs. IX, 263 pages

178 **Capillary Surfaces**
Shape – Stability – Dynamics, in Particular Under Weightlessness
By Dieter Langbein 2002. 182 figs. XVIII, 364 pages

179 **Anomalous X-ray Scattering for Materials Characterization**
Atomic-Scale Structure Determination
By Yoshio Waseda 2002. 132 figs. XIV, 214 pages

180 **Coverings of Discrete Quasiperiodic Sets**
Theory and Applications to Quasicrystals
Edited by P. Kramer and Z. Papadopolos 2002. 128 figs., XIV, 274 pages

181 **Emulsion Science**
Basic Principles. An Overview
By J. Bibette, F. Leal-Calderon, V. Schmitt, and P. Poulin 2002. 50 figs., IX, 140 pages

182 **Transmission Electron Microscopy of Semiconductor Nanostructures**
An Analysis of Composition and Strain State
By A. Rosenauer 2003. 136 figs., XII, 238 pages

183 **Transverse Patterns in Nonlinear Optical Resonators**
By K. Staliūnas, V. J. Sánchez-Morcillo 2003. 132 figs., XII, 226 pages

184 **Statistical Physics and Economics**
Concepts, Tools and Applications
By M. Schulz 2003. 54 figs., XII, 244 pages

185 **Electronic Defect States in Alkali Halides**
Effects of Interaction with Molecular Ions
By V. Dierolf 2003. 80 figs., XII, 196 pages

186 **Electron-Beam Interactions with Solids**
Application of the Monte Carlo Method to Electron Scattering Problems
By M. Dapor 2003. 27 figs., X, 110 pages

187 **High-Field Transport in Semiconductor Superlattices**
By K. Leo 2003. 164 figs.,XIV, 240 pages

188 **Transverse Pattern Formation in Photorefractive Optics**
By C. Denz, M. Schwab, and C. Weilnau 2003. 143 figs., XVIII, 331 pages

189 **Spatio-Temporal Dynamics and Quantum Fluctuations in Semiconductor Lasers**
By O. Hess, E. Gehrig 2003. 91 figs., XIV, 232 pages

190 **Neutrino Mass**
Edited by G. Altarelli, K. Winter 2003. 118 figs., XII, 248 pages

191 **Spin-orbit Coupling Effects in Two-dimensional Electron and Hole Systems**
By R. Winkler 2003. 66 figs., XII, 224 pages

192 **Electronic Quantum Transport in Mesoscopic Semiconductor Structures**
By T. Ihn 2003. 90 figs., XII, 280 pages

193 **Spinning Particles –Semiclassics and Spectral Statistics**
By S. Keppeler 2003. 15 figs., X, 190 pages

194 **Light Emitting Silicon for Microphotonics**
By S. Ossicini, L. Pavesi, and F. Priolo 2003. 206 figs., XII, 284 pages

195 **Uncovering *CP* Violation**
Experimental Clarification in the Neutral K Meson and B Meson Systems
By K. Kleinknecht 2003. 67 figs., XII, 144 pages

196 **Ising-type Antiferromagnets**
Model Systems in Statistical Physics and in the Magnetism of Exchange Bias
By C. Binek 2003. 52 figs., X, 120 pages

Springer Tracts in Modern Physics

197 **Electroweak Processes in External Electromagnetic Fields**
By A. Kuznetsov and N. Mikheev 2003. 24 figs., XII, 136 pages

198 **Electroweak Symmetry Breaking**
The Bottom-Up Approach
By W. Kilian 2003. 25 figs., X, 128 pages

199 **X-Ray Diffuse Scattering from Self-Organized Mesoscopic Semiconductor Structures**
By M. Schmidbauer 2003. 102 figs., X, 204 pages

200 **Compton Scattering**
Investigating the Structure of the Nucleon with Real Photons
By F. Wissmann 2003. 68 figs., VIII, 142 pages

201 **Heavy Quark Effective Theory**
By A. Grozin 2004. 72 figs., X, 213 pages

202 **Theory of Unconventional Superconductors**
By D. Manske 2004. 84 figs., XII, 228 pages

203 **Effective Field Theories in Flavour Physics**
By T. Mannel 2004. 29 figs., VIII, 175 pages

204 **Stopping of Heavy Ions**
By P. Sigmund 2004. 43 figs., XIV, 157 pages

205 **Three-Dimensional X-Ray Diffraction Microscopy**
Mapping Polycrystals and Their Dynamics
By H. Poulsen 2004. 49 figs., XI, 154 pages

206 **Ultrathin Metal Films**
Magnetic and Structural Properties
By M. Wuttig and X. Liu 2004. 234 figs., XII, 375 pages

207 **Dynamics of Spatio-Temporal Cellular Structures**
Henri Benard Centenary Review
Edited by I. Mutabazi, J.E. Wesfreid, and E. Guyon 2005. approx. 50 figs., 150 pages

208 **Nuclear Condensed Matter Physics with Synchrotron Radiation**
Basic Principles, Methodology and Applications
By R. Röhlsberger 2004. 152 figs., XVI, 318 pages

209 **Infrared Ellipsometry on Semiconductor Layer Structures**
Phonons, Plasmons, and Polaritons
By M. Schubert 2004. 77 figs., XI, 193 pages

210 **Cosmology**
By D.-E. Liebscher 2005. Approx. 100 figs., 300 pages

211 **Evaluating Feynman Integrals**
By V.A. Smirnov 2004. 48 figs., IX, 247 pages

213 **Parametric X-ray Radiation in Crystals**
By V.G. Baryshevsky, I.D. Feranchuk, and A.P. Ulyanenkov 2006. 63 figs., IX, 172 pages

214 **Unconventional Superconductors**
Experimental Investigation of the Order-Parameter Symmetry
By G. Goll 2006. 67 figs., XII, 172 pages

215 **Control Theory in Physics and other Fields of Science**
Concepts, Tools, and Applications
By M. Schulz 2006. 46 figs., X, 294 pages

216 **Theory of the Muon Anomalous Magnetic Moment**
By K. Melnikov, A. Vainshtein 2006. 33 figs., XII, 176 pages

217 **The Flow Equation Approach to Many-Particle Systems**
By S. Kehrein 2006. 24 figs., XII, 170 pages

219 **Inelastic Light Scattering of Semiconductor Nanostructures**
By Christian Schüller 2007. 105 figs., XII, 178 pages

FSC
www.fsc.org
MIX
Papier | Fördert
gute Waldnutzung
FSC® C083411

Kolibri 360 GmbH
Ferdinand-Jühlke-Straße 7
99095 Erfurt, Deutschland
produktsicherheit@kolibri360.de